Monographs in Electrochemistry

Surprisingly, a large number of important topics in electrochemistry are not covered by up-to-date monographs and series on the market, some topics are even not covered at all. The series "Monographs in Electrochemistry" fills this gap by publishing in-depth monographs written by experienced and distinguished electrochemists, covering both theory and applications. The focus is set on existing as well as emerging methods for researchers, engineers, and practitioners active in the many and often interdisciplinary fields, where electrochemistry plays a key role. These fields range – among others – from analytical and environmental sciences to sensors, materials sciences and biochemical research.

Dmitriy Pruttskov · Aleksander Andriiko ·
Aleksei Kirichenko

Electrolytic Production of Al–Si Alloys

Theory and Technology

Springer

Dmitriy Pruttskov
Department of Metallurgical Technologies
Ecology and Technogenic Safety
Zaporizhzhya National University
Zaporizhzhia, Ukraine

Aleksander Andriiko
Department of Chemical Technology
National Technical University of Ukraine
"Igor Sikorsky Kyiv Polytechnic Institute"
Kyiv, Ukraine

Aleksei Kirichenko
Department of Metallurgical Technologies
Ecology and Technogenic Safety
Zaporizhzhya National University
Zaporizhzhia, Ukraine

ISSN 1865-1836 ISSN 1865-1844 (electronic)
Monographs in Electrochemistry
ISBN 978-3-031-29251-4 ISBN 978-3-031-29249-1 (eBook)
https://doi.org/10.1007/978-3-031-29249-1

This Springer imprint is published by the registered company Springer Nature Switzerland AG
The registered company address is: Gewerbestrasse 11, 6330 Cham, Switzerland

Preface

Alloys based on the Al–Si system (usually containing from 6 to 13 mass% of silicon) have successfully replaced steel and cast iron in aviation and mechanical engineering. The reasons are not only their lower density but the unique combination of high casting, mechanical and operational characteristics as well. The major part of the production of these alloys is carried out exclusively by dissolving pieces of metallurgical grade silicon (MG) in molten aluminum in electric furnaces with forced stirring (synthetic method). About 2.3 million tons of MG silicon were produced in ore-thermal furnaces in 2018. One-third of this quantity was used in the production of alloys. Charcoal is mainly used as a reducing agent in the production of MG silicon. Therefore, the countries with rapid natural reforestation (China, Brazil) have clear advantages. Modern powerful furnaces are equipped with carbon electrodes of large dimensions (e.g., for a three-phase 16.5 MW furnace, electrodes with a diameter of 1.2 m are used). Their production is carried out at specialized plants. The construction of these electrodes is performed by means of a nipple connection, which is a weak link; therefore, they frequently break, leading to emergencies.

Large amounts of crystalline MG silicon are produced by electrothermy to obtain alloys with aluminum. The aluminum is produced by electrolysis of cryolite–alumina melts. Therefore, the situation is somewhat paradoxical: In nature, oxides of aluminum and silicon are in paragenesis and colossal material resources are spent on their separation using a combination of complex hydro- and pyro-metallurgical processes to produce alumina and then they are combined again, but already as individual substances. It would be logical to carry out their joined reduction directly from natural low-iron raw materials (kaolin $Al_2O_3 \cdot 2SiO_2 \cdot 2H_2O$; disthene (kyanite), sillimanite $Al_2O_3 \cdot SiO_2$, etc.). There are two options for their reduction: a carbothermic process in an arc furnace and an electrolytic process in an aluminum electrolyzer. Both processes, as well as the synthetic method, were used at the aluminum plant in Zaporizhzhia city (Ukraine). The instrumental–technological scheme of these processes is shown in Fig. 1.

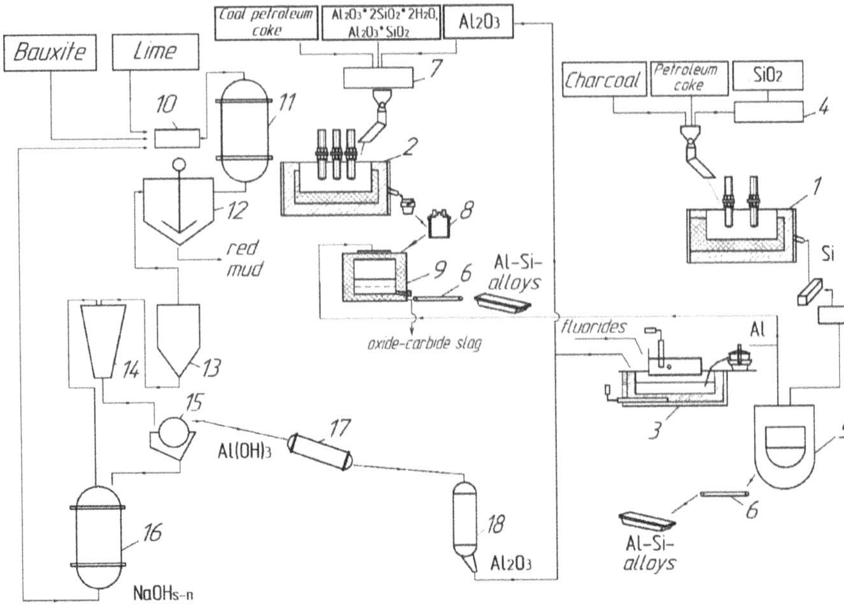

Fig. 1 Designations: (1) single-phase ore-thermal furnace; (2) three-phase ore-thermal furnace; (3) Söderberg electrolyzer; (4) coarse grinding; (5) electric furnace for producing Al–Si alloys; (6) casting conveyor; (7) briquetting; (8) refining; (9) dilution and filtration; (10) fine grinding; (11) leaching; (12) thickening and washing; (13) decomposition; (14) hydroseparation; (15) filtration; (16) evaporator; (17) calcination; (18) bunker

Industrial testing of the technology for obtaining Al–Si alloys in ore-thermal furnaces using briquettes of alumina, aluminosilicates (kaolin, disthene-sillimanite concentrate) and ground reducing agent (petroleum coke, coal) has proved the possibility of producing a "primary alloy" of composition (in mass%): Al:58–59; Si:36–37; Fe: ~1.6; Ti:~0.6; Ca:~1.0; SiC:~2.0. After refining, it was diluted with aluminum to a concentration of 11 to 13 mass% Si.

The technology using single-phase working electrode ore-thermal furnaces with conductive bottom as a second electrode (Miguet) and a capacity of 10 MW was successfully operated in the 1930s with high technical and economic efficiency. The design of the furnace made it possible to localize the thermal energy released by the electric arc directly under the electrode, that is, in the zone where the redox reactions take place.

Then, starting from the 1960s, three-phase furnaces with a capacity of 16.5 MW have been introduced. However, an unforeseen problem occurred: Due to the so-called batch conductivity, a noticeable part of the current passed directly between the electrodes, and not through the bottom, causing some heat dissipation. As a result, this process was accompanied by the formation of an oxide–carbide slag (up to 20 mass%), the removal of which is a complex engineering task, which naturally worsened its technical and economic performance [1, 2].

Thus, the production of commercial alloys (6 to 13 mass% Si) can be carried out (both by carbothermal and synthetic methods) only with the simultaneous use of ore-thermal furnaces and electrolyzers. Therefore, it seems tempting to carry out a one-stage process for obtaining alloys of this concentration by joint reduction of Al_2O_3 and SiO_2 directly in the electrolytic cell. Such direct method based on the electrolysis of a cryolite-oxide melt is slag-free.

Tests of the technology started in the 1930s in the Söderberg electrolyzer with a current load of 23 kA [3]. Alloys containing 10.6 to 13.0 mass% Si have been obtained. However, over time, the normal course of the unit was disturbed: An increase in temperature and the appearance of sediments were noted. Then, it was not possible to find out the causes of detrimental phenomena and eliminate them.

After the appearance of reports on the production of Al–Si alloys in industrial electrolyzers [4, 5], these developments were resumed. The results obtained by the present authors were scattered in a series of Russian language publications, most of which (especially between 1986 and 1997), unfortunately, were not available to interested Western readers. In our opinion, this information is still of scientific value, and therefore, the idea arose to systematize it and present it in this book. This scientific and technical research field has not lost its relevance, as shown by modern publications on this problem. They are critically analyzed from the point of view of the results obtained by the authors, and prospects for the development of technology are outlined.

Zaporizhzhia, Ukraine — Dmitriy Pruttskov
Kyiv, Ukraine — Aleksander Andriiko
Zaporizhzhia, Ukraine — Aleksei Kirichenko

References

1. Belyaev A, Rappoport M, Firsanova L (1953) Elektrometallurgiya aliuminiya (electrometallurgy of aluminum). Metalurgizdat, Moscow, p 720 (In Russian)
2. Ragulina R, Emlin B (1972) Elektrotermiya kremniya i silumina (electrothermy of silicon and silumin). Metallurgiya, Moscow, p 240 (In Russian)
3. Belyaev A (1944) Metallurgiya legkikh metallov (metallurgy of light metals). Metalurgizdat, Moscow, p 543 (In Russian)
4. Hannach R, Osborne J, Templeton G, Frazer E, Welch B (1977) Molten salts electrolysis in metal production. Proc Int Symp Grenoble 7–13
5. Tabereax A, McMinn C (1978) TMS light metals, pp 209–222. Republised as Tabereaux AT, McMinn CJ (2016) Production of aluminum-silicon alloys from sand and clay in hall cells. In: Bearne G, Dupuis M, Tarcy G (eds) Essential readings in light metals. Springer, Cham. https://doi.org/10.1007/978-3-319-48156-2_158

Acknowledgements

We worked on preparation of this book in Ukraine in hard times of war. Evidently, bombs, rockets and shells from above were not favorable for such kind of work. However, we received strong moral support both in Ukraine and abroad. In particular, we cordially thank German scientists—Lothar Griesser, Achim Lunk and Fritz Scholz—for their kind letters with words of support and encouragement. We hope that this book will be useful for the researchers (we know many of them) who works in field of electrochemistry of cryolite melts.

We are greatly indebted to our Editors Fritz Scholz and Laszlo Peter—for their hard work on the improvement of the manuscript's quality. Special thanks for their efforts, which cannot be overestimated.

Kyiv–Zaporizhzhia
June 2022

Contents

About the Authors

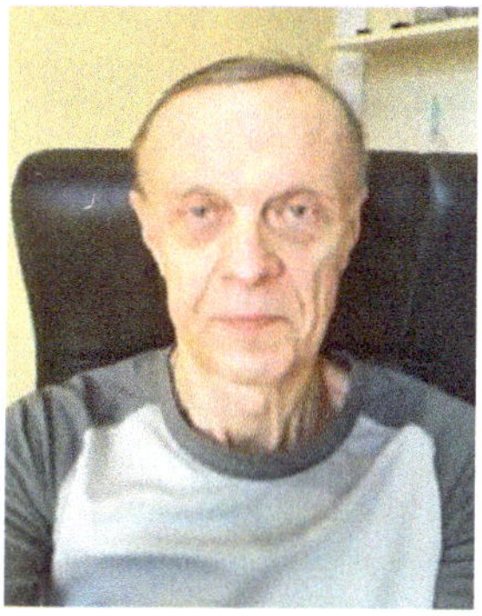

Dmitriy Pruttskov is a professor at Zaporizhzhia National University, Ukraine. He graduated from Zaporizhzhia Industrial Academy in 1978 and obtained his diploma in metallurgical engineering. After that, he worked at All-Union Research and Design Institute of Titanium (Zaporizhzhia, Ukraine) where he held the positions of engineer, junior, senior researcher, and head of the laboratory of electrometallurgical processes. He received his degree of Ph.D. (Kandidate) in 1984 from the Institute of General and Inorganic Chemistry (IGICH) of the National Academy of Science of Ukraine and his D.Sc. in 1995 from the Institute of High-Temperature Electrochemistry of Russian Academy of Science (Ekaterinburg, Russia) for the works in the field of electrochemistry of cryolite melts. His research interests are focused on the physical chemistry and technology of magnesium, aluminum, silicon, titanium, and their alloys, as well as refractory inorganic compounds. The results of his research activities were published in 150 scientific papers. Also, he is a co-author of 30 inventions, most of which are used in the metallurgical industry.

Aleksander Andriiko had been the head chair of General and Inorganic Chemistry of the National Technical University of Ukraine "Kyiv Polytechnic Institute" (NTUU "KPI") since 2003 till 2021. After retirement from this position, he remains a professor of this chair (part-time). He obtained his diploma in Chemical Engineering in 1974 and worked in the Institute of General and Inorganic Chemistry (IGICH) of the National Academy of Science of Ukraine since then and till 2003. He received his degrees of Ph.D. (Kandidate) in 1981 and D.Sc. in 1993 from IGICH for the works in the field of electrochemistry of molten fluorides. He worked as a scientific advisor in the Republic of Korea (1998) and as a visiting professor at the Centre of Electrochemical Surface Technology (CEST) in Wiener Neuschtadt, Austria (Summer 2008). He has carried out researches in various fields of inorganic chemistry (coordination chemistry of 3D metals, high-temperature chemistry of molten fluorides), physical chemistry (phase equilibria in molten salt systems), electrochemistry (thermodynamics of metal–electrolyte interface, electrodeposition of refractory and other metals from molten salts, lithium batteries and active materials for them) and, more recently, nanochemistry of inorganic oxide materials. The results of his research activities were published in more than 120 papers in refereed journals and monograph *Many-Electron Electrochemical Processes* (Springer 2013). Also he is the co-author of 24 inventions and the author of the chapter in: *Electrochemistry in a Divided World* / F. Sholz Ed. (Springer 2015).

Aleksei Kirichenko is an associate professor, Candidate of Technical Sciences, and an associate professor of the Department of Metallurgical Technologies, Ecology and Technogenic Safety of the Zaporizhzhia National University (Zaporizhzhia, Ukraine). In 2000, he graduated from the Zaporizhzhia State Engineering Academy (Zaporizhzhia, Ukraine) and received the qualification of a metallurgical engineer. In the same year, he started working at the department of ferrous metallurgy of the Zaporizhzhia State Engineering Academy. In 2012, he received a Ph.D. (candidate) degree from Zaporizhzhia State Engineering Academy for his research work on the utilization of red sludges with the production of dispersed metal-carbon materials. In 2014, he received the academic title of an associate professor of the Department of Metallurgy of Ferrous Metals. In the period 2019–2022, he held the position of head of the Department of Metallurgy of the Engineering Educational and Scientific Institute of the Zaporizhzhia National University (Zaporizhzhia, Ukraine). His main interest is the processing of secondary metallurgical waste, metal-carbon materials, and electrochemistry. The list of his scientific publications includes more than 100 scientific works, three co-authored books and six patents for inventions.

About the Series Editors

Fritz Scholz is an emeritus professor at the University of Greifswald, Germany. Following studies of chemistry at Humboldt University, Berlin, he obtained a Dr. rer. nat. and a Dr. sc. nat. (habilitation) from that University. In 1987 and 1989, he worked with Alan Bond in Australia. His main interest is in electrochemistry and electroanalysis. He has published more than 300 scientific papers, and he is the editor and co-author of the book *Electroanalytical Methods* (Springer, 2002, 2005, 2010, and Russian Edition: BINOM, 2006), the co-author of the book *Electrochemistry of Immobilized Particles and Droplets* (Springer 2005), the co-editor of the *Electrochemical Dictionary* (Springer, 2008; 2nd ed. 2012), and the co-editor of volumes 7a and 7b of the *Encyclopedia of Electrochemistry* (Wiley-VCH 2006). In 1997, he has founded the *Journal of Solid State Electrochemistry* (Springer) and served as the editor-in-chief until 2021. In 2014, he has founded the journal *ChemTexts—The Textbook Journal of Chemistry* (Springer). He is the editor of the series *Monographs in Electrochemistry* (Springer) in which modern topics of electrochemistry are presented. He introduced the technique "Voltammetry of Immobilized Microparticles" for studying the electrochemistry of solid compounds and materials, he introduced three-phase electrodes to determine the Gibbs energies of ion transfer between immiscible liquids, and he has extensively studied the interaction of free oxygen radicals with metal surfaces, as well as the interaction of liposomes with the surface of mercury electrodes in order to assess membrane properties. He is also interested in the history of science and is the co-editor of the English translation of *Wilhelm Ostwald's Autobiography* (Springer 2017) and the editor of the book *Electrochemistry in a Divided World: Innovations in Eastern Europe in the 20th Century* (Springer 2015).

László Péter is a scientific advisor at the Wigner Research Centre for Physics (Budapest, Hungary). He graduated as a teacher of physics and chemistry at the Eötvös University of Budapest in 1992 and obtained the Ph.D. degree at the same university in 1995. After spending 2 years in the USA as a postdoctoral fellow and 1 year in Japan as a visitor scientist, he returned to his home country and started working in the predecessor of the Wigner Research Centre. His main interest is electrochemistry; in particular, the formation of solid phases in electrochemical processes, the physical properties of electrodeposited materials and the electrochemical background of corrosion. Beside electrochemistry, he deals with various fields of experimental physical chemistry and research aspects of industrial problems. His publication list includes more than 100 research papers, two chapters and one monograph. He is the founding secretary of the conference series called *International Workshops on Electrodeposited Nanostructures (EDNANO)*. In 2013, he became the doctor of the Hungarian Academy of Sciences. From 2020, he is one of the topical editors of *Journal of Solid State Electrochemistry* (Springer).

Chapter 1
Exchange Reactions in Na_3AlF_6–Al_2O_3–SiO_2 Melts

1.1 Introduction

This chapter presents the results of studies of physico-chemical transformations occurring in the Na_3AlF_6–Al_2O_3–SiO_2 melt. Apparently, the interaction of SiO_2 with Na_3AlF_6 melt was studied in [1] for the first time. The authors noted the release of a gaseous substance with a pungent odor, which allowed them to assume the following reaction:

$$4Na_3AlF_6 + 3SiO_2 \rightarrow 12NaF + 2Al_2O_3 + 3SiF_4\uparrow \quad (1.1)$$

According to (1.1), complete removal of silicon from the melt should take place. However, a year later, in [2] it was reported that the mass loss of the Na_3AlF_6–SiO_2 melt decreases with time, and it is less than expected according to reaction (1.1). The authors attributed this discrepancy to the formation of aluminosilicate (metakaolin) in the melt, which is no longer decomposed by cryolite.

$$4Na_3AlF_6 + 7SiO_2 \rightarrow 2(Al_2O_3 \cdot 2SiO_2) + 12NaF + 3SiF_4\uparrow. \quad (1.2)$$

Then, in [3], it was found that when Al_2O_3 is added to the Na_3AlF_6–SiO_2 binary melt, silicon losses are practically eliminated, and this fact was confirmed later in [4]. In [5], nepheline, $NaAlSiO_4$, was found in quenched samples of the ternary melt. It was assumed that it is the formation of nepheline (rather than metakaolin [2]) that eliminates the volatilization of silicon compounds.

The authors of [6] proposed another formula of the resulting aluminosilicate, namely jadeite $NaAlSi_2O_6$. These authors treated the solidified Na_3AlF_6–SiO_2 mixture with an $AlCl_3$ solution. They believed that fluorides would dissolve, while oxides would remain in the precipitate, and by analyzing the latter one could determine the composition of the aluminosilicate. However, due to the hydrolysis of $AlCl_3$, the pH of the medium was less than 7, and therefore, partial decomposition of nepheline took place [7]. The latter got its name from the Greek word νεφέλωμα

D. Pruttskov et al., *Electrolytic Production of Al–Si Alloys*,
Monographs in Electrochemistry, https://doi.org/10.1007/978-3-031-29249-1_1

(cloud, nebula), because when it is exposed to an HCl solution, a silicic acid gel forms, resembling a cloud or nebula:

$$NaAlSiO_{4(solid)} + 4HCl_{(solution)} \rightarrow NaCl_{(solution)} + AlCl_{3(solution)} + H_4SiO_{4(solid)}. \quad (1.3)$$

Obviously, this is the reason why the authors found an excess of SiO_2 in the precipitate relative to the nepheline stoichiometry.

Thus, the available information was scarce and contradictory; therefore, a comprehensive study of ligand exchange reactions in the Na_3AlF_6–Al_2O_3–SiO_2 melt was carried out [8–10].

1.2 Products of Transformations in the Na_3AlF_6–Al_2O_3–SiO_2 Melts

The melt evaporation rate was determined by measuring the weight loss of a Pt crucible with a mixture of Na_3AlF_6 and SiO_2 using an E-2D1 mechanoelectric transducer. Figure 1.1 shows the obtained gravimetric curves.

As expected, additions of SiO_2 lead to an increase of the mass loss rate of the mixtures in comparison with pure Na_3AlF_6. However, overtime, the evaporation rate of the mixtures decreases, approaching that for pure salt, although, according to

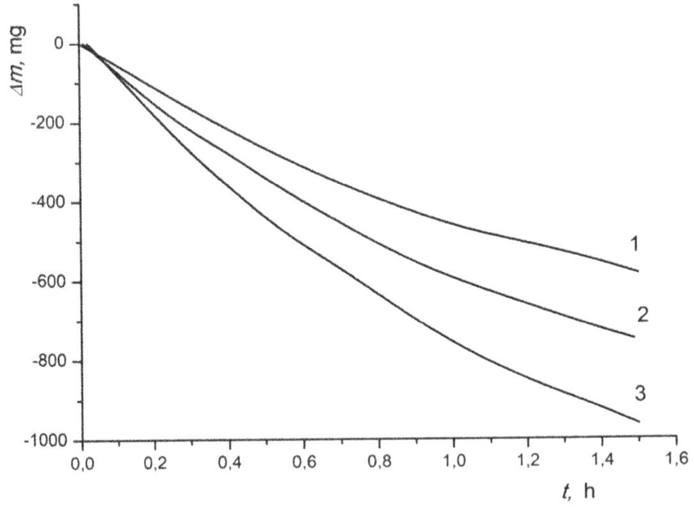

Fig. 1.1 Thermogravigrams of Na_3AlF_6–SiO_2 melts. $T = 1323$ K; evaporation surface area 3.8 cm^2; the initial mass of the sample is 8.0 g. Designations: (1) cryolite; (2), (3) cryolite containing 2.5 and 5 mass% SiO_2, respectively

chemical analysis data, less than half of the added SiO_2 is lost (which is consistent with [2].

It should be noted that the samples of SiO_2 dissolve in Na_3AlF_6 in a few minutes, while an increased rate of weight loss of the mixture is observed within 60–90 min. Hence, it can be concluded that the exchange reactions leading to the evaporation of silicon compounds are homogeneous and proceed in the volume of the melt and not only at the moment of dissolution of the oxide in the salt.

The gaseous reaction products were collected by passing argon through a sealed steel retort in which there was a glassy carbon crucible with a Na_3AlF_6 (6 mass%)–SiO_2 melt, followed by freezing the gas mixture with liquid nitrogen and fractional distillation.

In the mass spectrum obtained on the MI-1201 instrument, a particle was found with a ratio of mass number to charge $m/e = 104$, which corresponds to SiF_4. Its fragmentation products were also detected (Fig. 1.2). The mass spectrum of SiF_4 obtained according to the procedure [11] turned out to be identical, and therefore, it can be considered as the main gaseous product of the exchange reaction.

This melt was kept at $T = 1323$ K for 1.5 h and quenched. Its thin section was studied on MBI-6 and MIM-8 optical microscopes and by microprobe analysis (X-ray microspectral method) using an MS-46 instrument. The existence of glasses (Fig. 1.3) with a refractive index $N = 1.50$–1.51 and composition, mass%: Na: 15.6 $\pm$ 2.1; Al: 18.2 $\pm$ 2.6; Si: 20.7 $\pm$ 2.2 was established. Both the composition and the N value of the glass [12] correspond to nepheline $NaAlSiO_4$.

Therefore, the reaction can be written in molecular form as follows:

$$n\mathrm{Na_3AlF_6} + 2\mathrm{SiO_2} \rightarrow (n-1)\mathrm{Na_3AlF_6} + \mathrm{NaAlSiO_4} + \mathrm{SiF_4}\uparrow + 2\mathrm{NaF}. \quad (1.4)$$

In the presence of Al_2O_3, silicon losses are eliminated [4, 5], and the discovery of nepheline [6] makes it possible to represent this transformation as

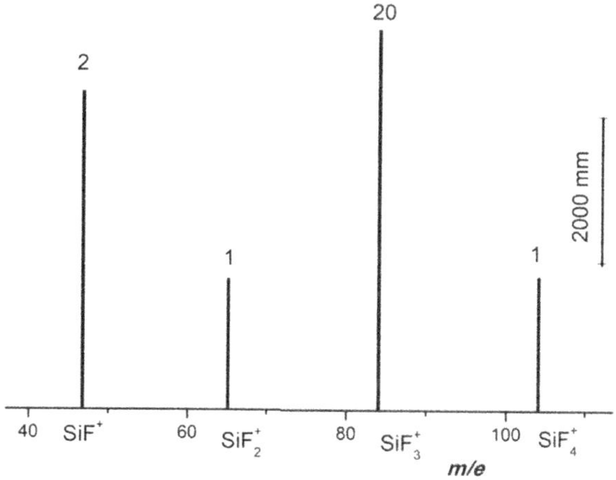

Fig. 1.2 Mass spectrum of the gaseous compound above the Na_3AlF_6–SiO_2 melt. The numbers near the lines indicate the multiplicity of their intensity decrease

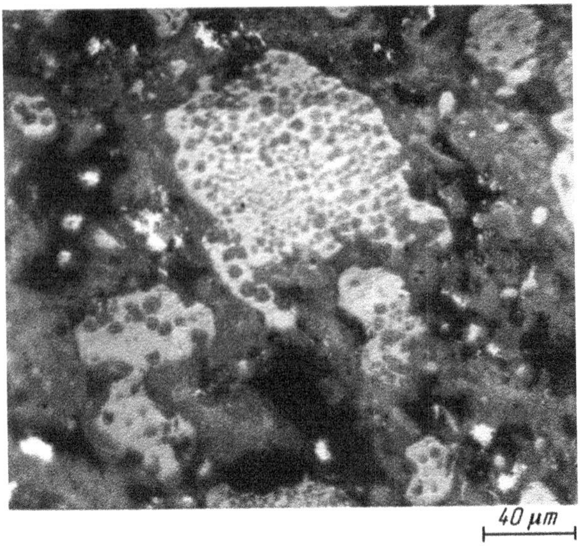

Fig. 1.3 Microstructure of the quenched Na_3AlF_6 (6 mass%)–SiO_2 melt. Light inclusions—nepheline glass, dark spots on the glass background—fluorides (magnification 400)

$$n\mathrm{Na_3AlF_6} + 3\mathrm{SiO_2} + 2\mathrm{Al_2O_3} \rightarrow (n-1)\mathrm{Na_3AlF_6} + 3\mathrm{NaAlSiO_4} + 2\mathrm{AlF_3}. \quad (1.5)$$

X-ray phase analysis, carried out on a DRON-0.5 diffractometer, showed the presence of villiomite NaF in the products of reaction (1.4), and chiolite $Na_5Al_3F_{14}$ in reaction (1.5), which, according to the phase diagram [13–15], indicate the appearance of an excess of AlF_3 in the melt with respect to the stoichiometry of Na_3AlF_6.

Obviously, the transformations occurring in the Na_3AlF_6 melt, containing both SiO_2 and Al_2O_3 together, for the studied dilute solutions (as is the case in industrial electrolysis) end in the formation of stable nepheline-like particles. At the same time, with a significant content of SiO_2 in the fluoride-oxide mixture, the formation of albite $NaAlSi_3O_8$ is possible [16, 17].

1.3 Criterion of Thermal Stability of the Na_3AlF_6–Al_2O_3–SiO_2 Melts

We can assume that the probability of transfer of SiF_4 into the gas phase depends on the ratio of Al_2O_3/SiO_2 oxides dissolved in cryolite. To check this assumption, a tensimetric study of molten mixtures was performed using the thermographic version of the boiling point method [18, 19]. The scheme of the setup is shown in Fig. 1.4. The molten mixture is in a boron nitride crucible (1), and the crucible itself is placed in a graphite cup (2) with a tightly ground lid, through which a differential thermocouple (3) was introduced into the working space of the cup using two-channel corundum tubes, the cold ends of which were placed in a Dewar vessel (5) with melting ice. A

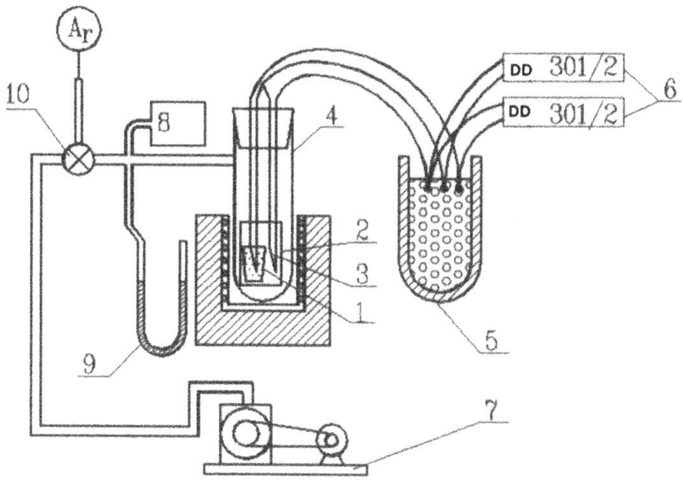

Fig. 1.4 Scheme of installation for measuring vapor pressure. Designations: (1) crucible with melt; (2) graphite beaker with a tightly ground lid; (3) differential thermocouple; (4) quartz retort; (5) Dewar vessel; (6) combined digital device DD-301-2; (7) vacuum pump VM-461; (8) receiver; (9) pressure gauge; (10) valve

hole in the lid of the beaker with a diameter of 0.5 mm served as a diaphragm limiting the thermal diffusion transfer of the substance. The entire system was placed in a quartz retort (4), which was closed with a rubber stopper. The retort was connected to a vacuum system consisting of pump (7) of the VM-461M type, a receiver (8) with a capacity of 5 l and a mercury manometer (9). The air in the cell was replaced by argon after it was evacuated using a three-way valve (10). The temperature of the melt and the readings of the differential thermocouple were measured by two combined digital devices (6) DD301/2.

The retort filled with argon was placed in a heated furnace. After the preset temperature was reached, the external pressure was gradually lowered, which was controlled by the manometer. At the moment when the vapor pressure of the studied mixture and the external pressure became equal, the evaporation rate increased sharply and, therefore, the melt temperature somewhat decreased, which was recorded by a differential thermocouple. According to the obtained results, at a constant content of SiO_2 and increasing the content of Al_2O_3 in the melt, the vapor pressure gradually decreased (Fig. 1.5).

One can observe that when the ratio of atoms in the melts O/Si = 4 (as in nepheline), the vapor pressure only slightly exceeds that for stoichiometric Na_3AlF_6. Most likely, this deviation is due to the appearance of an excess of AlF_3 in the melt according to reaction (1.5), since more "acidic" melts (with cryolite ratio c.r. < 3) have a higher vapor pressure [13–15]. Further increase of O/Si ratio results in a smooth decrease of pressure, similarly to the binary Na_3AlF_6–Al_2O_3 melt [13–15]. Obviously, the parameter O/Si $\geq$ 4 can be considered as the criterion for the thermal stability of the Na_3AlF_6–Al_2O_3–SiO_2 ternary melt.

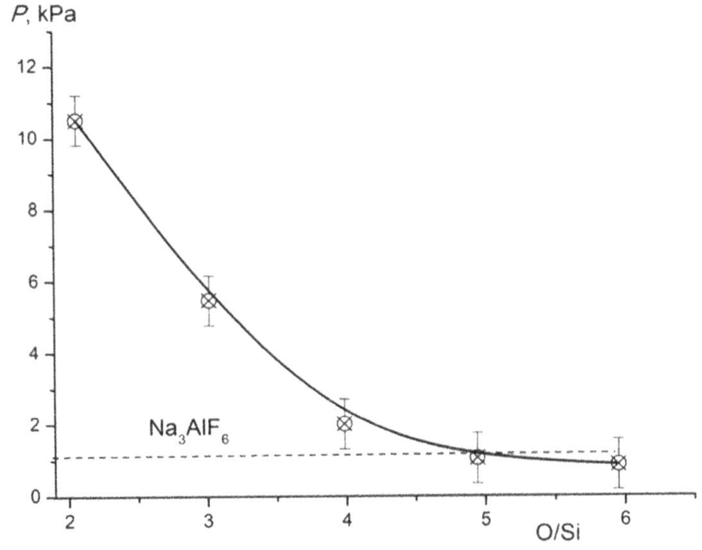

Fig. 1.5 Dependence of vapor pressure of Na_3AlF_6–Al_2O_3–SiO_2 melt on the O/Si ratio. The content of SiO_2 is constant (5 mass%); at O/Si = 2, the content of Al_2O_3 is 0 mass%; at 3: 3 mass%; at 4: 5.5 mass%; at 5: 8.5 mass%; at 6: 10 mass%. $T = 1323$ K

1.4 Ionic Structure of the Na_3AlF_6–Al_2O_3–SiO_2 Melts

Let us present the transformations in ionic form. Cryolite in the molten state dissociates according to the scheme [13–15]:

$$Na_3AlF_6 \leftrightarrows 3Na^+ + AlF_6^{3-} \leftrightarrows 3Na^+ + AlF_4^- + 2F^-. \quad (1.6)$$

When SiO_2 is dissolved, its silicon–oxygen framework is destroyed. According to cryoscopic measurements, dilute solutions of SiO_2 in the Na_3AlF_6 melt are characterized by the formation of new species, and this condition corresponds to the formation of such a tetrahedron [20]

$$SiO_2 + 2F^- \rightarrow SiO_2F_2^{2-}. \quad (1.7)$$

The existence of such an anion is also admitted in more recent works [21, 22]. Then, obviously, the formation of SiF_4 occurs by the exchange reaction

$$2SiO_2F_2^{2-} + AlF_6^{3-}(AlF_4^-) \rightarrow AlSiO_4^- + SiF_4\uparrow + 6(4)F^-. \quad (1.8)$$

The volatilization kinetics of SiF_4 was studied by isothermal exposure of the Na_3AlF_6–SiO_2 melt in glassy carbon crucible and subsequent analysis of the quenched samples for silicon content. An example of the obtained dependence,

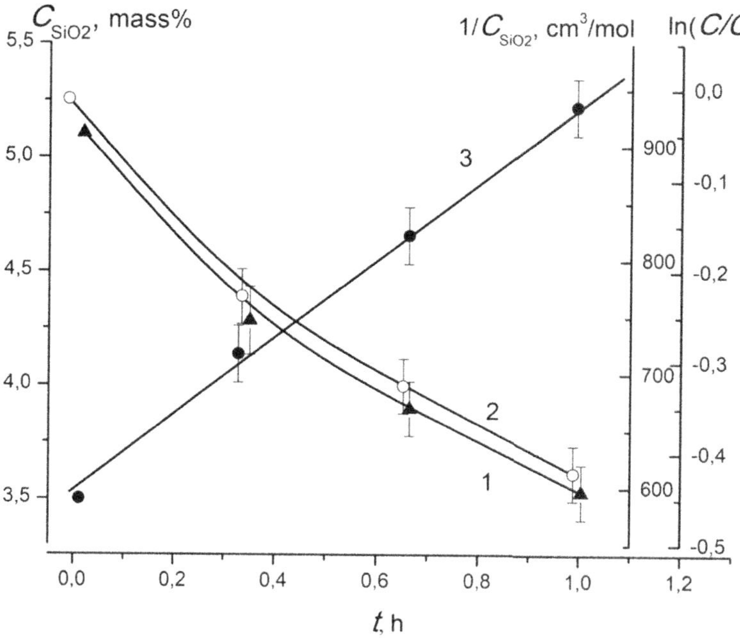

Fig. 1.6 Dynamics of decrease of SiO_2 content in cryolite and its processing by kinetic equations. Designations: (1) dependence C_{SiO2}–t; (2), (3) curve 1, presented in the coordinates ln C/C_0–t and $1/C_0$–t, respectively

corrected by taking into account the rate of evaporation of the cryolite itself, is shown in Fig. 1.6.

The figure also presents the processing of curve 1 according to the equations of formal kinetics, which showed that the curve is linearized in the coordinates $1/C$–τ. This indicates that reaction (1.8) is approximated by a second-order equation [23].

Using Arrhenius plot, its activation energy is found to be 90.3 ± 4.5 kJ/mol. The obtained data allow us to conclude that the homogeneous reaction (1.8) proceeds according to the associative mechanism [24].

The transformation mechanism should be different in the presence of Al_2O_3. Its dissolution is also accompanied by the formation of oxyfluoride ions, for example, [20]

$$AlF_6{}^{3-} + Al_2O_3 \rightarrow 3AlOF_2{}^{-}. \tag{1.9}$$

Although, according to modern studies, the charged particles with an Al–O–Al bond may form [13–15], this is not of significant importance for our reasoning. Then it can be assumed that with the simultaneous additions of Al_2O_3 and SiO_2 into cryolite, nepheline-like ions are formed as a result of the polycondensation reaction

$$SiO_2F_2{}^{2-} + 2AlOF_2{}^{-} \rightarrow AlSiO_4{}^{-} + AlF_6{}^{3-} \rightarrow AlSiO_4{}^{-} + 2F^{-} + AlF_4{}^{-}. \tag{1.10}$$

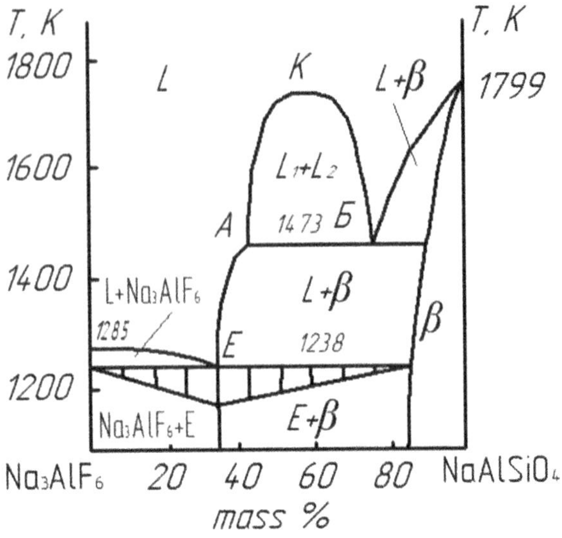

Fig. 1.7 Phase diagram of Na_3AlF_6–$NaAlSiO_4$ system

Assuming the existence of other complex aluminum ions [13–15], we can write similar equations. The absence of loss of SiF_4 from the melt containing excess of Al_2O_3 [4, 5] (Fig. 1.5) allows us to conclude that the rate of reaction (1.10) is significantly higher than that for (1.8).

Certain information about the structure of the melt can be obtained from the Na_3AlF_6–$NaAlSiO_4$ phase diagram. It was built using visual-polythermal, differential-thermal (Q-1500 device), and quenching methods (Fig. 1.7).

The cryolite side is of simple eutectic type with eutectic point at 32 mass% and $T = 1238$ K. On the $NaAlSiO_4$ side, there is a segregation dome, inside which two immiscible liquids coexist: a solution of nepheline in cryolite L1 (its composition varies along the limiting saturation line AK) and a solution of cryolite in nepheline L2 (line BK). At a temperature of ~ 1473 K, a monotectic transformation occurs, when a solid β-solution is precipitating from the solution during cooling into a solid phase, and the liquid phase corresponds to composition A. Further lowering of the temperature leads to a change in the composition of the liquid along the liquidus line AE with the release of an excess of solid β-solution.

Segregation of this melt was also observed in [25]. This system was later studied in [16]. The authors recorded a similar flat shape of the liquidus line from the side of fluoride, which they explained as positive deviations from the ideal solution, but segregation phenomena were not established.

Summarizing, we can conclude that the phase diagram of Na_3AlF_6–$NaAlSiO_4$ is a stable pseudobinary section of the quaternary reciprocal system Na, Al, Si//O and F. Obviously, in melts with compositions inside the polyhedron limited by the lines Na_2O–$NaAlSiO_4$–Al_2O_3–AlF_3–NaF, ligand exchange reactions leading to the formation of SiF_4 do not take place, since the condition $O/Si > 4$ is satisfied (Fig. 1.8). Suitable for the electrolytic production of Al–Si alloys are compositions lying on the plane of the triangle Na_3AlF_6–$NaAlSiO_4$–Al_2O_3, in its lower corner. Since electrolysis is carried out from cryolite melts with an excess of AlF_3, the position of the triangle may change, which is indicated in Fig. 1.8 with dash-dotted line.

For electrolysis technology, the fluoride part of the diagram is more important.

Approximation of the initial section of the liquidus line by the cryoscopic equation showed that the number of new particles introduced into the melt is about three times less than follows from the stoichiometric formula of nepheline $NaAlSiO_4$. The latter belongs to the class of framework silicates and has a tridymite-like structure composed of six [AlO_4] and [SiO_4] tetrahedra (three of each) alternating according to the Lowenstein rule with vertices directed in opposite sides and enclosed in a ring (Fig. 1.9) [7, 26].

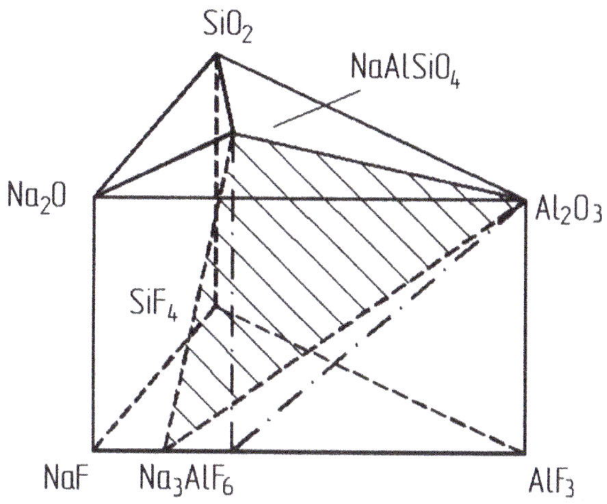

Fig. 1.8 Region of thermal stability of Na, Al, Si//O, F system. See explanations in the text

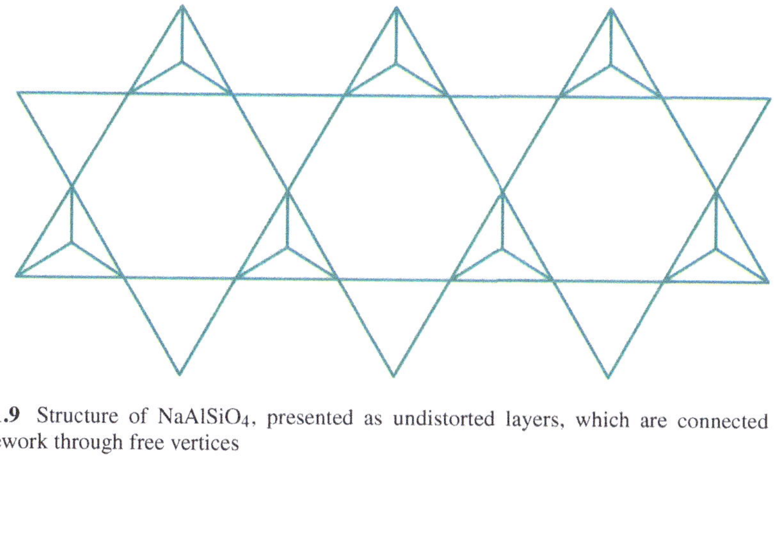

Fig. 1.9 Structure of $NaAlSiO_4$, presented as undistorted layers, which are connected into a framework through free vertices

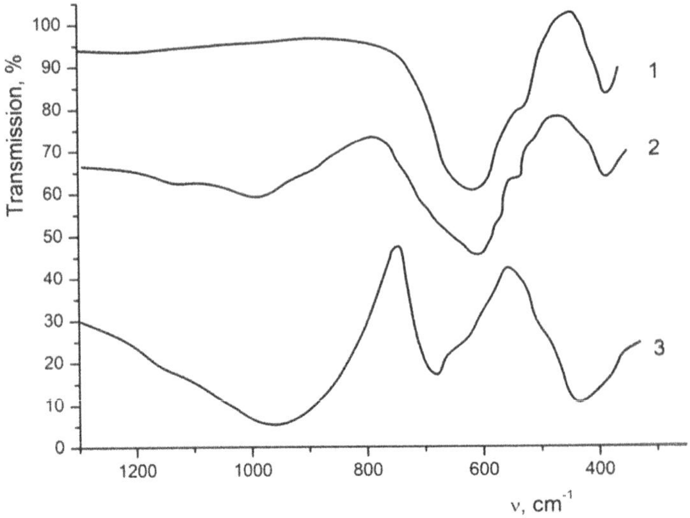

Fig. 1.10 Infrared spectra of initial ingredients and quenched Na_3AlF_6 (10 mass%)–$NaAlSiO_4$ melt. Designations: (1) cryolite; (2) quenched mixture and (3) nepheline

Therefore, it was proposed to write its crystal chemical formulas $[NaAlSiO_4]^{3\infty}$ [27]. Obviously, when it is dissolved in cryolite, the newly formed ions retain their cyclicity. This conclusion is supported by the IR spectrum of the quenched Na_3AlF_6–$NaAlSiO_4$ melt (device A-302, KBr tablets 1:100), which had the same absorption bands as the original ingredients (Fig. 1.10).

All oxygen atoms in nepheline are bridging O^0 [7, 26]. It is also known that the binding energy of the Si–O pair is higher than that of Al–O [28, 29]. When nepheline is dissolved in cryolite, the frame is destroyed, accompanied by the destruction of part of the bridges (Al–O–Si) (e.g., in the middle cycle, Fig. 1.9), silicon remains coordinated by oxygen ($Si–O^-$) (i.e., oxygen becomes terminal), and then aluminum must bind fluorine melt ion ($Al–F^-$). This conclusion is consistent with the polarity rule [25], according to which the equilibria of exchange reactions in halide–oxide melts are shifted toward the realization of maximally and minimally polar bonds, i.e., a more acidic cation is combined with a more basic anion and vice versa, which corresponds in our case to the Si–O and Al–F bonds. Therefore, the structure of the polyanion can be represented as follows (Fig. 1.11). Most likely, the Al–F bond, both in cryolite and in dissolved nepheline, absorbs electromagnetic radiation at close frequencies, and therefore, no additional bands appear in the IR spectrum (Fig. 1.10).

Then, the dissolution process can be represented by the reaction

$$3[NaAlSiO_4]_{solid} + n\{3Na^+ + AlF_6{}^{3-}\}_{melt} \rightarrow \left\{3(n+3)Na^+ + \left[(SiO_4)_3(AlF_2)\right]_{complex}{}^{9-} + (n-3)AlF_6{}^{3-} + 3AlF_4{}^-\right\}_{melt} \quad (1.11)$$

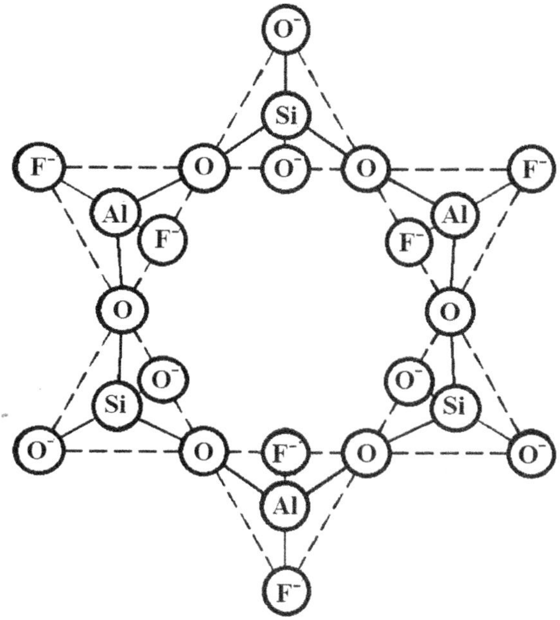

Fig. 1.11 Structure of a hypothetical polyanion

Recent studies of fluoride–aluminosilicate systems by NMR, Raman and IR spectroscopy confirm this interpretation [30]. Accordingly, Eqs. (1.8) and (1.10) should be represented as follows:

$$\{6SiO_2F_2{}^{2-} + 3AlF_6{}^{3-}(AlF_4{}^-)\}_{melt} \rightarrow \left\{[(SiO_4)_3(AlF_2)_3]_{complex}{}^{9-} + 12(6)F^-\right\}_{melt} + 3SiF_4\uparrow \quad (1.12)$$

$$\{3SiO_2F_2{}^{2-} + 6AlOF_2{}^-\}_{melt} \rightarrow \left\{[(SiO_4)_3(AlF_2)_3]_{complex}{}^{9-} + 3AlF_4{}^-\right\}_{melt} \quad (1.13)$$

Hence, it is possible to formulate the physical meaning of the criterion of thermal stability O/Si $\geq$ 4 for Na_3AlF_6–Al_2O_3–SiO_2 melt. Obviously, its oxide component tends to acquire such a configuration in which silicon cations are completely screened by oxygen anions from fluorine ions of the melt, and this condition corresponds to the formation of ring groups. Based on this concept, two-ligand tetrahedral $SiO_2F_2{}^{2-}$ are considered only as the intermediates, which agrees with [25]. They are consumed by reactions (1.12) and (1.13), and the Si–F bond is not realized in nepheline-like cycles.

A similar explanation for the absence of SiF_4 losses from $MF(MF_2)$–$M_2O(MO)$–SiO_2 melts with high basicity ($M_2O(MO)/SiO_2 \geq 1$) was proposed earlier in [25, 31, 32]. The formation of strong silicon-oxygen cyclic polyanions is allowed, or with an even greater basicity ($M_2O(MO)/SiO_2 \geq 2$) with the existence of anion $SiO_4{}^{4-}$ as the limiting option.

In view of the above, it seems that the Na_3AlF_6–$NaAlSi_3O_8$ system [16, 17, 33] is not thermally stable (i.e., not in thermodynamic equilibrium), since O/Si = 8/3, i.e., < 4 (Fig. 1.5). The above is also true for the cryolite–metasilicate system, where the ratio is [34]. It means that an exchange reaction should proceed with the evolution of SiF_4 into the gas phase, and, with time, the ionic-molecular composition of the oxide part of the melt will come to nepheline-like cycles.

Finally, it should be noted that our conclusions regarding the form of silicon-containing complex ions in melts of the Na_3AlF_6–Al_2O_3–SiO_2 system can be applied only to dilute solutions. Increasing the silicon content in the melt leads to the association of nepheline-like cycles with the formation of larger framework structures through the implementation of the (Si–O–Al) bond already between neighboring six-nuclear species, and then the Al–F bond will form only at surface aluminum atoms. It can be assumed that at concentrations of nepheline > 32 mass%, the cluster sizes reach a critical value and, since the aluminosilicate microphase has a higher surface tension compared to the fluoride component of the melt [25], it collects into drops forming a segregation dome of two immiscible liquids on the phase diagram (Fig. 1.7).

1.5 Conclusions

The physico-chemical mechanism of transformations occurring in the Na_3AlF_6–Al_2O_3–SiO_2 melt is as follows. The dissolution of the initial oxides in cryolite leads to the destruction of their crystal lattices and the formation of the corresponding oxyfluoride ions. Then, nepheline-like cyclic polyanions are formed in the melted mixture by two competing reactions: exchange interaction between $SiO_2F_2^{2-}$ and AlF_6^{3-} (AlF_4^-) with release of SiF_4 into the gas phase and polycondensation interaction between $SiO_2F_2^{2-}$ and $AlOF_2^-$ (or complex ions with a different composition [13–15]) without the formation of volatile silicon compounds. The rate of the polycondensation reaction is significantly higher than that of the exchange reaction. As a result, the formation of SiF_4 does not occur when oxides Al_2O_3 and SiO_2, taken in the required proportion, are loaded into the cryolite melt.

The criterion for the thermal stability of the Na_3AlF_6–Al_2O_3–SiO_2 melt is the atomic ratio O/Si $\geq$ 4, which corresponds to nepheline. Under the conditions of industrial electrolysis, the content of Al_2O_3 in cryolite is within 2–8 mass%, while the concentration of SiO_2 (to ensure a stable technological regime) should not exceed 0.5–0.6 mass%, which means that the condition O/Si $\geq$ 4 is constantly met, and hence, the formation of SiF_4 is minimized.

In a cryolite melt, nepheline-like patterns retain the cyclicity inherent to its crystal lattice. The formula of hypothetical polyanions can be represented as $[(SiO_4)_3(AlF_2)_3]^{9-}$ with no Si–F bonds in the structure. The existence of such cycles must be taken into account when studying the process of electrochemical production of silicon.

References

1. Batashev K, Zhurin A (1933) Metallurg 2:60–72 (in Russian)
2. Mashovets V, Artobolevskaya E (1934) Proc NIIS Aluminium Leningrad GONTI 7:20–26 (in Russian)
3. Antipin P, Ivanova L (1950) Tsvetnye Metally (Non-ferrous Met) 2:49–53 (in Russian)
4. Fellner P, Matiasovsky K (1973) Chem Zvesty 27:737–741
5. Ivanova L (1960) Izv VUZov Khim Khim Tekhnol (Chem Chem Technol) 6:970–974 (in Russian)
6. Snow R, Welch B (1972) Proc Aust Inst Min Metall 241:81–86
7. Deer W, Howie R, Zussman J (1963) Rock-forming minerals, vol 4: framework silicates. Longman, London, 435 p
8. Pruttskov D, Andriiko A, Titaev P (1989) Ukr Chem J 55:569–574 (in Russian)
9. Pruttskov D, Titaev P, Andriiko A, Pirozhkova V (1990) Ukr Chem J 56:470–475 (in Russian)
10. Pruttskov D, Krivoruchko N (1997) Rasplavy (Melts) 11:81–89 (in Russian)
11. Rapoport F, Il'inskaya A (1963) Laboratornye metody polucheniya chistykh gazov (Laboratory methods for preparation of pure gases). Goskhimizdat, Moscow, 420 p (in Russian)
12. Winchell A (1964) The microscopical characters of artificial inorganic solid substances: optical properties of artificial minerals. Academic Press, New York, 439 p
13. Grjotheim K, Krohn C, Malinovsky M, Matiasovsky K, Thonstad J (1982) Aluminium electrolysis, 2nd edn. Aluminium-Verlag, Dusseldorf, 442
14. Grjotheim K, Krohn C, Malinovsky M, Matiasovsky K, Thonstad J (1977) Aluminium electrolysis. Aluminium-Verlag, Dusseldorf, 350 p
15. Thonstad J, Fellner P, Haarberg GM, Hives J, Kvande H, Sterten A (2001) Aluminium electrolysis, 3rd edn. Aluminium-Verlag, Dusseldorf, 359 p
16. Rutlin J, Grande T (1999) J Am Ceram Soc 82:2539–2544
17. Simko F, Rakhmatullin A, Korenko M, Bessada C (2021) J Mol Liq 328:115453
18. Novikov G, Polyachenok O (1961) Russ J Inorg Chem 6:1951–1952 (in Russian)
19. Zharskiy I, Novikov G (1988) Fizicheskiye metody v khimii (Physical methods in chemistry). Vysshaya Shkola, Moscow, 271 p (in Russian)
20. Antipin L, Vazhenin S (1964) Elektrokhimiya rasplavlennykh solei (Electrochemistry of molten salts). Metallurgizdat, Moscow, 355 p (in Russian)
21. Dolejs D, Baker D (2005) Geochim Cosmochim Acta 69:5537–5556
22. Feng Y, Li M, Hou W, Cheng B, Wang J, Li H (2021) ACS Omega 6:3745–3751
23. Schmid R, Sapunov V (1982) Non-formal kinetics: in search for chemical reaction pathways. Verlag Chemie Gmbh, Weinheim, 260 p
24. Basolo F, Pearson R (1967) Mechanisms of inorganic reactions: study of metal complexes in solutions. Wiley, New York, 701 p
25. Kogarko L, Kriegman L (1981) Ftor v silikatnykh rasplavakh i magmakh (Fluorine in silicate melts and magmas). Nauka, Moscow, 127 p (in Russian)
26. Libau F (1985) Structural chemistry of silicates. Springer-Verlag, Berlin, 347 p
27. Bokiy G (1971) Kristallokhimiya (Crystal chemistry). Science, Moscow, 400 p (in Russian)
28. Appen A (1970) Khimiya stekla (Glass chemistry). Khimiya, Moscow, 351 p (in Russian)
29. Epellbaum M (1980) Silikatnye rasplavy s letuchimi komponentami (Silicate melts with volatile components). Nauka, Moscow, 256 p (in Russian)
30. Mysen B, Richet P (2018) Silicate glasses and melts, 2nd edn. Elsevier Science, 720 p
31. Kozakevitch P (1954) Rev Met 51:588–594
32. Baak T, Olander A (1955) Acta Chem Scand 9:1350–1354
33. Siljan O-J, Grande T, Shoning C (2001) Aluminium 71(4):294–300
34. Shimko F, Boca M (2007) Helv Chim Acta 90:1529–1537

Chapter 2
Kinetics and Mechanism of Si(IV) Electroreduction in Na_3AlF_6–Al_2O_3–SiO_2 Melts on Al Cathode

2.1 Introduction

Silicon is not reduced from aqueous solutions because of the very negative potential of the Si(IV)/Si redox system, and therefore, hydrogen is released at the cathode [1, 2]. However, it is possible to obtain both elemental silicon and its alloys by electrolysis of melts. It seems that this was done for the first time by Minet [3], who carried out the electrolysis of the NaCl–Na_3AlF_6–Al_2O_3–SiO_2 melt and obtained an Al (8.9 mass%)–Si alloy. In the USSR, studies of the electrochemical properties of silicon were carried out in oxide melts by Esin and his co-workers [4, 5] and in halide molten electrolytes, by Delimarskiy and his team [6, 7]. The ideas about its stepwise reduction were developed in these works.

Silicon has an outer electronic configuration $3s^23p^2$, and the ionization potentials of its p-electrons (8.149 and 16.34 eV) are much lower than those of s-electrons (33.46 and 45.13 eV) [8]. Therefore, in compounds with non-metals, it possesses the valances two and four.

The properties of SiO monoxide have been studied in detail [9, 10]. It is stable at high temperatures and disproportionates when cooled

$$2SiO \rightarrow Si + SiO_2. \tag{2.1}$$

For example, in thin sections of solidified slag from the smelting of ferrosilicon, numerous inclusions of the smallest particles of elemental silicon were found, apparently formed by reaction (2.1). Such structures are considered to be the result of decomposition of compounds in lower oxidation states [11, 12].

Thus, the possibility of stepwise reduction of Si from the Na_3AlF_6–Al_2O_3–SiO_2 melt seemed to be reasonable, but there were no systematic data on its behavior in this melt. The decomposition voltage of SiO_2 in cryolite was measured in [13–15]. It turned out to be 0.3–0.4 V less than that for Al_2O_3, which allows us to consider silicon as a more "noble" element than aluminum. Chronopotentiometric study of the

D. Pruttskov et al., *Electrolytic Production of Al–Si Alloys*,
Monographs in Electrochemistry, https://doi.org/10.1007/978-3-031-29249-1_2

cathode reaction was carried out in [16]. However, the authors came to the conclusion that a single-step discharge occurs with a slow preceding reaction of the formation of electroactive particles.

There is also a discussion about the chemical nature of silicon reduction in an electrolytic cell: Does this process occur by aluminothermy, or does it obey the laws of joint discharge of ions? Here, the existing ideas about the mechanism of redox reactions involving liquid metal and electrolyte need to be considered. It is appropriate to mention the ideas about the mechanism of redox reactions involving liquid metal and electrolyte. When studying the decomposition of amalgams by acid solutions, it was found that this process cannot be explained from the standpoint of usual chemical kinetics [17]. To interpret this phenomenon, it was proposed in [18, 19] to consider it as two electrochemical reactions: metal oxidation and hydrogen reduction, occurring without the formation of local cathode and anode spots (for details, see [20]). These ideas were also applied to the reactions of reduction of metals from solutions using amalgams ("cementation") [21]. For redox reactions between liquid metals and molten electrolytes, this concept was developed in the works of Esin [4].

We have studied the features of Si(IV) reduction from a cryolite-oxide melt on an Al cathode in [22–29].

2.2 Reduction of Si(IV) by Liquid Al in Na_3AlF_6–Al_2O_3–SiO_2 Melts

Based on the idea that the extraction of elements from the melt by a more active liquid metal is the process of "cementation" [4] (aka: "contact exchange", "internal electrolysis"), it was interesting to measure the stationary potential of aluminum in the Na_3AlF_6–Al_2O_3–SiO_2 melt. The scheme of the used galvanic cell is shown in Fig. 2.1. In this series of experiments, an acidic fluoride melt (c.r. = 1.7) was used to reduce the temperature to $T = 1173$ K.

The cell potential difference (E) was measured with a high-resistance voltmeter of a P-5848 potentiostat with the registration of the time dependence (E–t) on a KSP-4 potentiometer.

Before the measurements, the composition of the melt in both compartments was the same, and therefore, E was zero. After adding SiO_2 to the outer crucible (which contained 40 g of fluorides and 10 g of aluminum), the electrode potential shifted sharply toward electropositive values, and the greater the SiO_2 addition, the faster the E growth and the higher its absolute values (Fig. 2.2).

With time, a sharp drop in E was observed (obviously, a decrease in the reaction rate) and it sets in the earlier, the higher the SiO_2 content is in the melt (curves 1 and 2). The broad maximum of curve 3 is due to the fact that, with the addition of SiO_2 in an amount of ≤ 0.5 mass%, there are no hindrances in redox reactions, and the electrode depolarization occurs due to the consumption of SiO_2 from the melt.

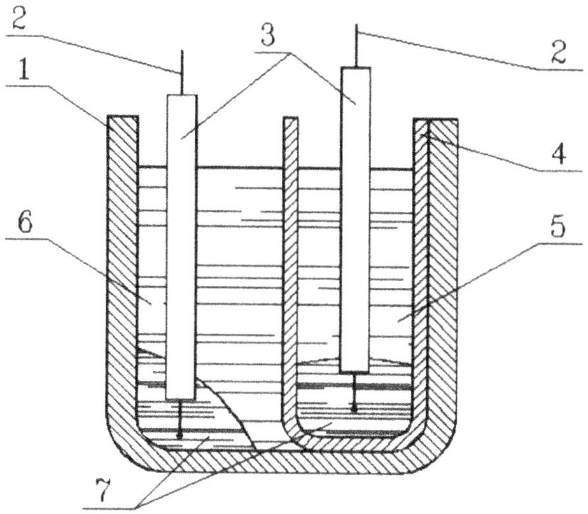

Fig. 2.1 Scheme of a galvanic cell for measuring the stationary potential of aluminum in the Na_3AlF_6–Al_2O_3–SiO_2 melt. Designations: (1) and (4) corundum crucibles; (2) tungsten current leads; (3) corundum tubes; (5) reference electrode Al | cryolite melt; (6) investigated electrode Al | cryolite melt + SiO_2; (7) aluminum

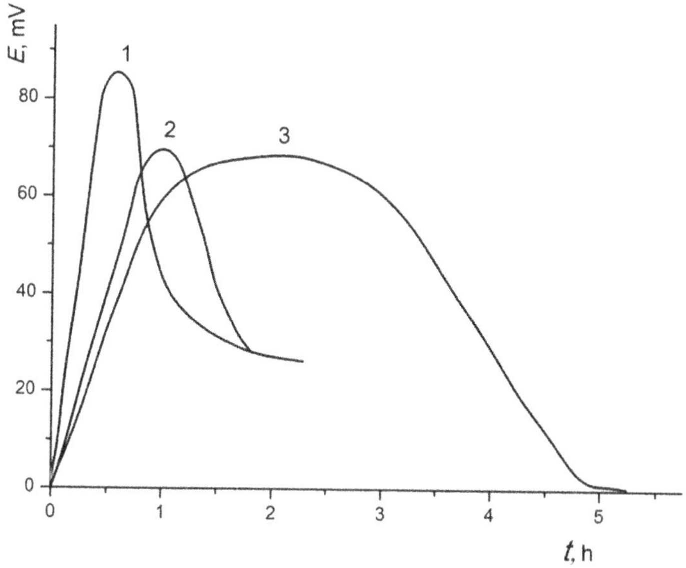

Fig. 2.2 Dynamics of changes in the stationary potential of aluminum depending on the content of SiO_2 in the melt. Designations: (1): 5 mass% SiO_2; (2): 1 mass% SiO_2; and (3): 0.5 mass% SiO_2

E tends to approach zero due to the low concentration of the formed alloy (< 0.1 mass% Si).

It means that the potential difference of the concentration cells

$$\text{Al} \mid \text{cryolite melt} \,\|\, \text{cryolite melt} + SiO_2 \mid \text{Al(Si)} \tag{2.2}$$

with diluted alloys (< 15 mass% Si) is negligible ($\leq$ 5 mV). Even at higher concentrations (~ 35 mass% Si), it is about 20 mV [30–32]. In our experiments, the maximum calculated alloy concentration could be only ~ 9 mass% (usually ~ 5 mass%) (curve 1 in Fig. 2.2), so the contribution of this circuit (2.2) to the measured E values (60–80 mV) can be neglected.

However, this conclusion is valid only in the absence of the "covering" effect caused by the slow withdrawal of Si from the surface into the bulk of liquid metal. To check it, the rapidly cooled crucible with the reaction mixture was cut on a Minosecar-2 machine, the metal was removed, and a section was prepared. This section was studied under an optical microscope (NU-2E) and micro-X-ray spectral method (MS-46 device). The structure of the alloy from the surface to the depth of the sample was uniform and consisted of light primary precipitates of the α-solid solution of Si in aluminum and acicular silumin eutectics (Fig. 2.3a).

Scanning with the probe raster also showed the absence of a concentration gradient (Fig. 2.3b), which means that the "covering" effect did not occur. This is explained by the fact that the diffusion coefficient of Si atoms in aluminum (46×10^{-5} cm^2/s, $T = 973$ K [33]) is almost an order of magnitude greater than that for Si ions in the cryolite-oxide melt (6.4×10^{-5} cm^2/s, $T = 1293$ K [15]).

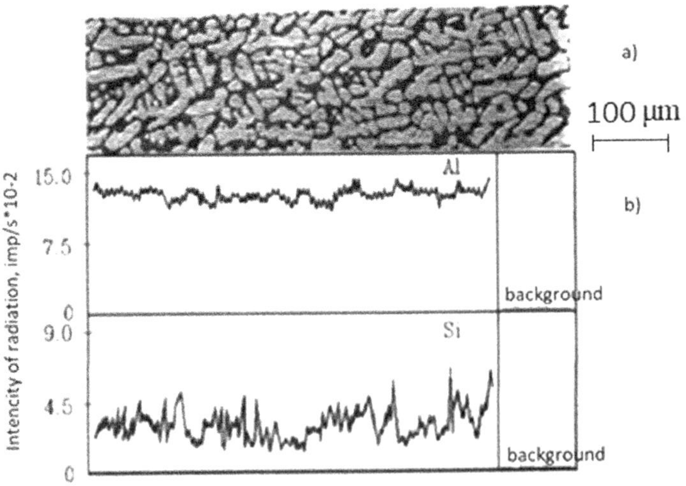

Fig. 2.3 Microstructure of the Al–Si alloy (**a**) and the concentration distribution curves of the intensity of $K\alpha$—radiation of the elements (**b**). The sample was obtained by exposing the melt containing 10 mass% SiO_2 onto the aluminum at $t = 1$ h. Magnification 100

Obviously, the shift of the aluminum potential to the anode side by 60–80 mV with the addition of SiO_2 indicates the occurrence of electrode reactions of its ionization

$$Al \rightarrow Al^{3+} + 3e^- \tag{2.3}$$

and discharge of silicon ions Si^{4+}

$$Si^{4+} + 4e^- \rightarrow Si(Al). \tag{2.4}$$

Such an interpretation of the aluminothermal reduction of Si from the Na_3AlF_6–Al_2O_3–SiO_2 melt provides the electrochemical background for the process of formation of the Al–Si alloy.

Additional information about the rate of this redox reaction was obtained by a chemical analytical method. A corundum crucible containing 5 g of aluminum and 30 g of cryolite was heated in a shaft furnace. Then, a weighed portion of SiO_2 was introduced, and after an exposure, the crucible was removed from the furnace and rapidly cooled.

The metal and salt were analyzed for Si content. With the addition of $SiO_2 \geq 1$ mass% after 0.5 h (which corresponds to a decrease in EMF on curves 1 and 2 in Fig. 2.2), the rate of transition of Si into the alloy decreases, but remains noticeable for 1.5 h (curve 1 in Fig. 2.4).

A sharp decrease in the content of SiO_2 in cryolite was noted, and it remained constant until the end of the experiment (curve 2 in Fig. 2.4). The total content of Si in the alloy and cryolite turned out to be less than what was introduced in the form of SiO_2 (i.e., a deficiency was observed, curve 3 in Fig. 2.4), and this difference

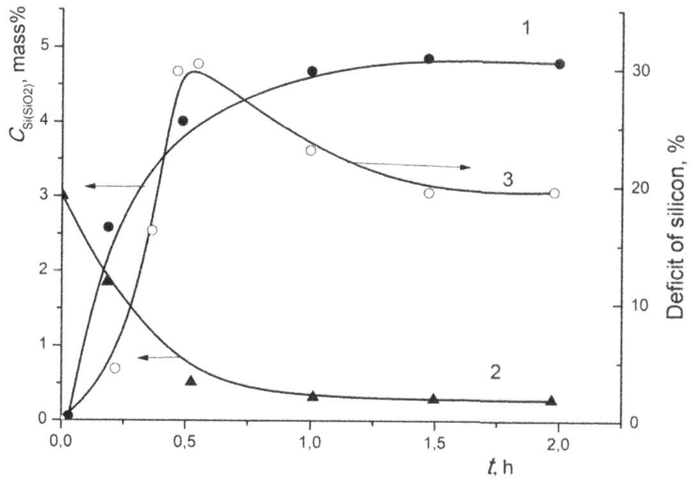

Fig. 2.4 Kinetics of aluminothermic reduction of silicon from cryolite-oxide melt. Designations: (1) concentration of Si in the alloy; (2) content of SiO_2 in cryolite; (3) silicon deficiency

strangely decreased with time (i.e., silicon seemed to disappear and then reappear) The anomalous course of this redox reaction was also noted in a recent work [34].

2.3 The Nature of the Passivation Layer on the Interphase Al | Na_3AlF_6–Al_2O_3–SiO_2

On a thin longitudinal section of the crucible at the phase boundary, a dark gray precipitate was found, easily separated from the metal and tightly linked to cryolite (Figs. 2.5 and 2.6).

Under an optical microscope (MBI-6, MIN-8), one can observe that the deposit consists of two layers. The upper, from the side of cryolite, layer "A", consists of metal and oxide-fluoride components. Its thickness is the greater, the higher is the amount of added SiO_2 and reaches 1.0–1.5 mm. According to microprobe analysis (MS-46 instrument), the metal is pure silicon. Also, amorphous SiO_2 with a refractive index $N = 1.458$ was found [35]. It is present in the form of thin veinlets in host fluorides. It is proved, however, that the product of interaction of initial SiO_2 with cryolite is nepheline $NaAlSiO_4$ (see Chap. 1). The presence of pure Si and amorphous SiO_2 together allows us to conclude that they were formed by a reaction similar to (2.1). As the distance from the phase boundary increases, their number decreases, and in the depth of the frozen cryolite, Si inclusions are fairly evenly distributed over the sample body (Fig. 2.7).

A significant content of decomposition products near the interface compared to the rest of the cryolite volume (Figs. 2.6 and 2.7) indicates that the Si(II) solubility limit has been exceeded and its excess precipitates into $[Si(II)]_{solid}$. During the electrolysis of andalusite (Al_2SiO_5) in a cryolite melt, the formation of a film containing silicon

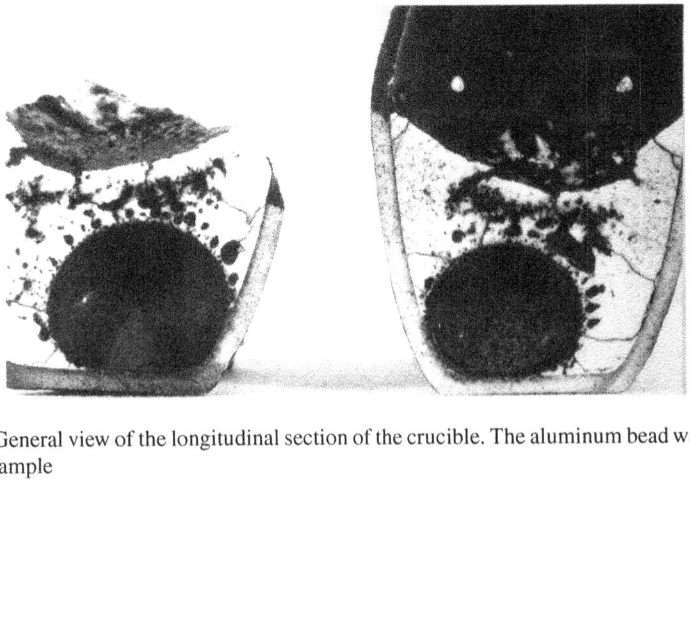

Fig. 2.5 General view of the longitudinal section of the crucible. The aluminum bead was removed from the sample

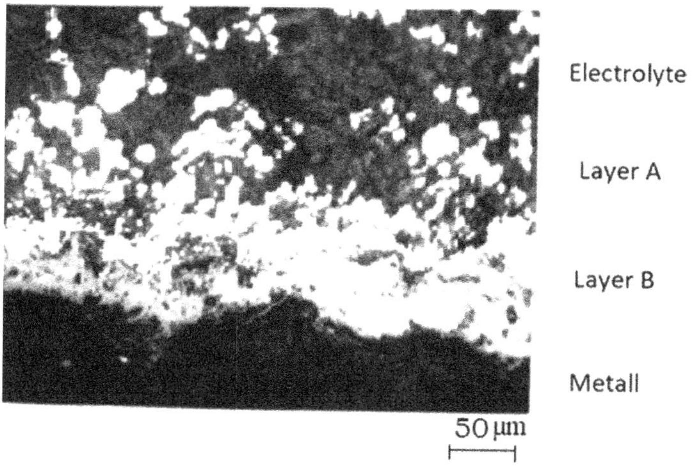

Fig. 2.6 Microstructure of the Al | Na_3AlF_6–Al_2O_3–SiO_2 interphase. The sample was obtained by exposing the melt containing 10 mass% SiO_2 and aluminum at $t = 1$ h. Magnification 210

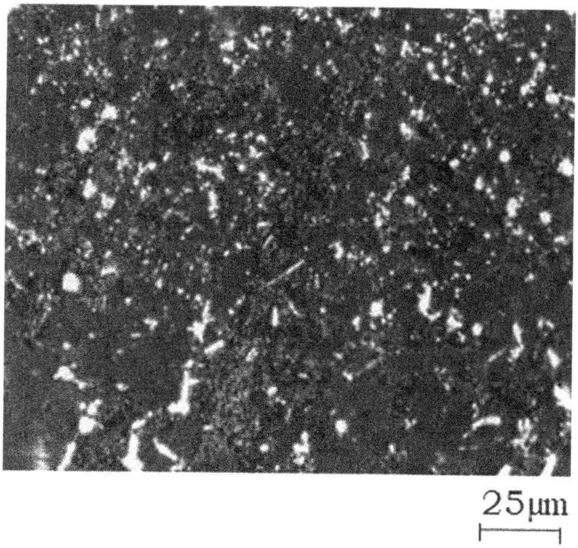

Fig. 2.7 Decomposition structure of the low valence silicon compound in the depth of the melt. Magnification 400

dendrites on the surface of the liquid cathode was also observed, but no explanation was given for this observation [36].

The thickness of the intermediate layer "B" does not depend on the added amount of SiO_2, and the main factor is the exposition time. The thickness is 0.010–0.015 mm for 0.5 h and 0.020–0.025 mm for 1 h. The optical constants of the crystals are changeable: in one case $Nq = 1.99$–2.10, which is close to 2.19 for Al_2O, and in the other $Nq = 1.74–1.75$, approaching 1.76, which corresponds to Al_2O_3 [36]. Microprobe analysis confirmed that these crystals are aluminum oxides.

These results permit to explain the behavior of the potentiometric and kinetic curves (Figs. 2.2 and 2.4). It can be argued that the reduction of the initial complex silicon ions conventionally designated as Si(IV) proceeds stepwise, and its intermediate compound Si(II) has a limited solubility in cryolite. Therefore, reaction (2.4) should be represented as follows:

$$\begin{array}{c} \mathrm{Si(IV)} \xrightarrow{2e^-} \mathrm{Si(II)} \xrightarrow{2e^-} \mathrm{Si(Al)} \\ \downarrow \\ [\mathrm{Si(II)}]_{\mathrm{solid}} \end{array} \tag{2.5}$$

Obviously, with the addition of $SiO_2 \leq 0.5$ mass%, the solubility limit of Si(II) is not attained, and therefore, blocking of the aluminum surface is not observed (curve 3, Fig. 2.2), whereas at high SiO_2 contents in cryolite (curves 1 and 2), the Si(II) solubility limit is exceeded and the $[Si(II)]_{solid}$ precipitate begins to form, which is localized at the phase boundary (Figs. 2.5 and 2.6). In our opinion, Si(II) should be considered as SiO monoxide solvated by fluorine ions of the cryolite melt.

The beginning of passivation corresponds to the falling portions of curves 1 and 2 and an inflection on the kinetic curve $C_{Si(Al)}–t$ (curve 1, Fig. 2.4). Further increase in the Si content in the alloy occurs due to the slow reaction of aluminum with the $[Si(II)]_{solid}$ deposit and the formation of Al_2O and Al_2O_3 oxides (layer "B" Fig. 2.6). Hence, the dependence of silicon deficiency on time becomes clear (curve 3 in Fig. 2.4).

Industrial tests of the method for obtaining aluminum alloys with other polyvalent metals by loading their oxides in an electrolyzer were accompanied by similar disorders [37–39]. Therefore, the question arose whether the observed phenomena are of common nature.

To test this idea, studies of the interphase boundary of the Al | $Na_3AlF_6–Al_2O_3–M_xO_y$ systems (where M is B, Ti, Zr, Hf, Mo, W) were carried out. In all cases, a passivating deposit similar to that shown in Figs. 2.5 and 2.6 was formed. Pure metal and its initial oxides (decomposition products) were found in layer "A", and layer "B" was represented by the same Al_2O and Al_2O_3. Hence, it follows that metal oxides in the low oxidation state have a lower solubility in the cryolite melt compared to the initial oxides. The chemical nature of this phenomenon is not yet clear.

2.4 Electroreduction of Si(IV) on Solid Electrodes

The redox reactions in the Al | $Na_3AlF_6–Al_2O_3–SiO_2$ system are associated with charge transfer across the phase boundary, and therefore, the classical methods of electrochemistry to determine the parameters of the cathode process are appropriate.

Chronopotentiometric measurements on a Ni cathode in the current density range 1.0–5.0 A/cm^2 were performed in [16]. Only one transition time (τ) was observed in the obtained E–t curves. The ($i\tau^{1/2}$)–i plot had a negative slope, and the E–t curves presented in semilogarithmic coordinates had a strong S-shaped distortion, which is typical for complex electrode processes [40]. The authors assumed that Si(IV) reduction proceeds in one four-electron step and is limited by the preceding chemical reaction. Since these data contradict the existence of the poorly soluble compound $[Si(II)]_{solid}$ (Figs. 2.5 and 2.6), additional chronopotentiometric measurements were performed.

Pt wire was used as both indicator and reference electrode. A glassy carbon crucible served as an auxiliary electrode and also as a container for the melt. The current source was a P-5848 potentiostat. The curves were recorded from the screen of an S1-48B oscillograph using a Zenit-3M SLR photographic camera. Current density range was $i = 0.05$–2.0 A/cm^2. The obtained E–t curves had from one to three inflections (Fig. 2.8).

The plots ($i\tau^{1/2}$)–i were not linear (Fig. 2.9), which indicates the presence of accompanying chemical reactions [40]. The analysis of the first wave corresponding to the diffusion-controlled process ($i\tau^{1/2} = \text{const}$) in the coordinates E versus $\ln(\tau^{1/2}-t^{1/2})/t^{1/2}$ showed that the number of electrons involved in this stage, as expected, equals to $n = 2$. Taking into account the existence of a passivating deposit (Figs. 2.5 and 2.6), the results obtained are interpreted as a stepwise discharge complicated by the inactivation of the intermediate (Eq. 2.5).

Later, the electroreduction of Si from a cryolite-oxide melt was studied in [41, 42]. The authors of [41] pointed out that Pt is not an indifferent metal and some of the transition times we observed (Fig. 2.8) could be associated with the formation of Pt_2Si. That is why they used a glassy carbon cathode as a more inert material. In the direct course of the voltammogram, one wave of Si reduction was observed, and

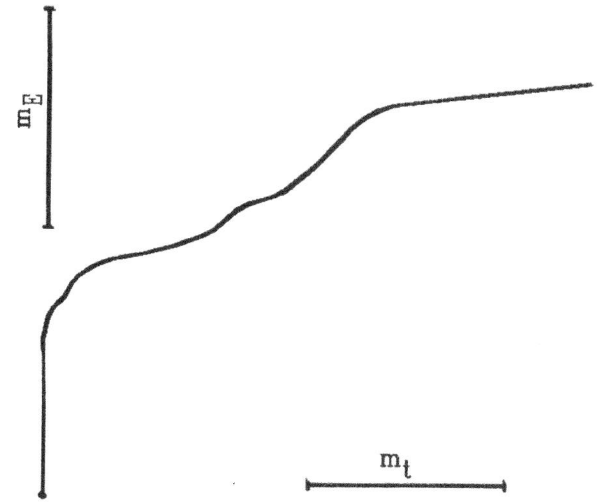

Fig. 2.8 Example of a chronopotentiometric curve obtained in the melt containing 0.75 mass% SiO_2. Current density $i = 0.335$ A/cm^2; scales: $m_E = 0.35$ V and $m_t = 0.06$ s

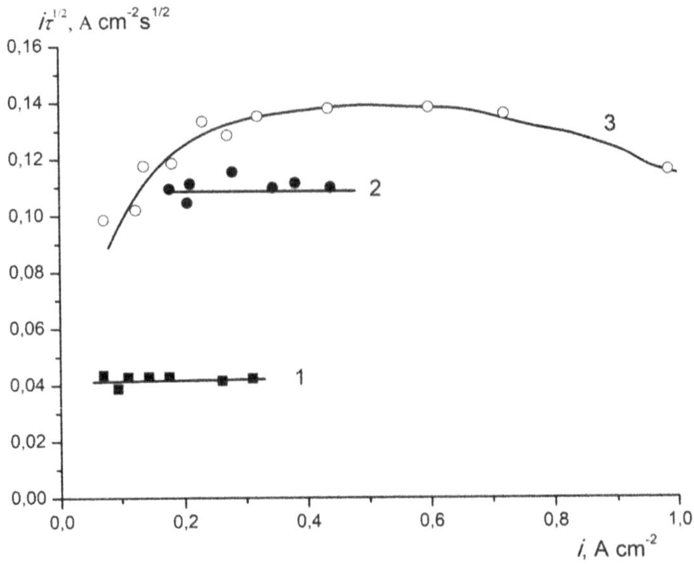

Fig. 2.9 Analysis of the chronopotentiometric transition times of the Si(IV) discharge at SiO_2 content of 0.75 mass%. The numbers correspond to the observed waves

during anodic polarization, two-stage dissolution of the product was detected. The plot of the cathode section of the *i–E* curve in the coordinates $\log[i/(i_d - i)] = f(E)$ showed a slope of 0.087–0.108 V, i.e., larger than for the reversible four-electron process ($2.3RT/4F = 0.058$ V). Based on these data, the authors concluded that Si(IV) is reduced irreversibly and in two steps.

In [42], other results were obtained using a graphite electrode. Both the stationary polarization curves and the reverse voltammograms showed a stepwise discharge of silicon. However, the anodic dissolution of the resulting product occurred in one stage, which was not explained.

The recent works where a cyclic chronovoltammetric study of the Na_3AlF_6–SiO_2 melt on graphite [43] as well as W and Ir [44] cathodes was performed should also be mentioned. A cathode peak with a weak "post-peak" and an extended peak of the oxidation of the electrolysis product were observed on the *i–E* curves obtained on graphite. Analyzing the effect of the SiO_2 content in cryolite and the potential sweep rate on the shape of the curves, the authors concluded that there is a stepwise reduction of silicon. Moreover, they believe that the Si(IV) → Si(II) step is limited by the delivery of the initial complexes to the electrode surface, while the rate of the Si(II) → Si(0) step is limited by charge transfer. When using Ir and W, both the direct and reverse branches of the *i–E* curves were observed to have two discharge and ionization peaks each. However, for Ir, in contrast to W, the peak potentials did not depend on the scan rate, and therefore, the authors assumed that the reduction of Si proceeds stepwise, but it is reversible on Ir and irreversible on W.

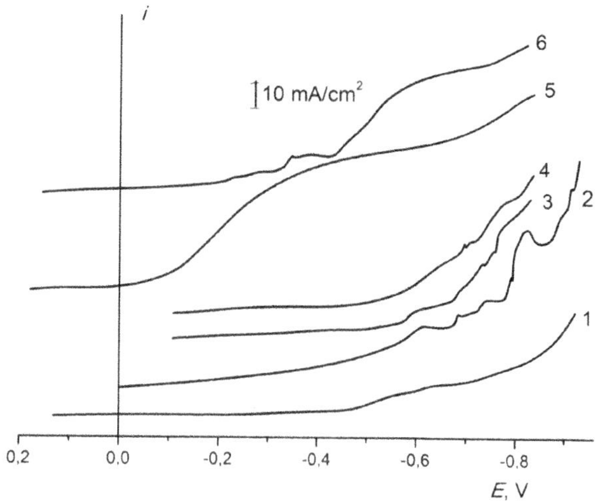

Fig. 2.10 Examples of voltammograms obtained in the Na_3AlF_6–$NaAlSiO_4$ melt under the conditions: $T = 1300$ K, scanning speed $V = 0.05$ V/s, $NaAlSiO_4$ concentration 1.38×10^{-4} mol/cm^3. Cathode material: (1) platinum; (2) spectral graphite; (3) molybdenum; (4) tungsten; (5) nickel; and (6) tantalum

We have noticed that the results of measurement depend on the material of the cathode. Therefore, for comparison, voltammograms were taken (PI-50-1 potentiostat, LKD4-003 recorder) on rods made of Pt, Ta, Ni, Mo, W, and spectrally pure graphite. The latter was also used as a reference electrode, while a glassy carbon crucible served as a container for the melt and as an auxiliary electrode. Remembering that the Na_3AlF_6–SiO_2 binary melt is thermally unstable, we introduced silica in form of nepheline $NaAlSiO_4$ (see Chap. 1). In this series of experiments, a fluoride melt of stoichiometric composition (c.r. = 3) was used. From one to four waves were observed on the obtained *i*–*E* curves (Fig. 2.10), which requires additional explanations. Therefore, the results of measurements on solid cathodes cannot be considered completely correct. More reliable information can be obtained using liquid electrodes.

2.5 Electrolytic Reduction of Si(IV) on a Liquid Electrodes

The subject of research should be a liquid electrode in which Si is electroactive, which means that its equilibrium potential is established; i.e., the system is electrochemically reversible. Such is the Cu–Si alloy. The values of Si activities in it are borrowed from [45, 46]. The alloys were obtained in a vacuum arc furnace from semiconductor Si and Cu grade MO (GOST 859). The Si potential versus the Al reference electrode in the Na_3AlF_6–$NaAlSiO_4$ melt was measured with a DD-301-2 instrument in a galvanic cell, the scheme of which is shown in Fig. 2.11.

Figure 2.12 shows the change of the Si potential depending on the experimental conditions.

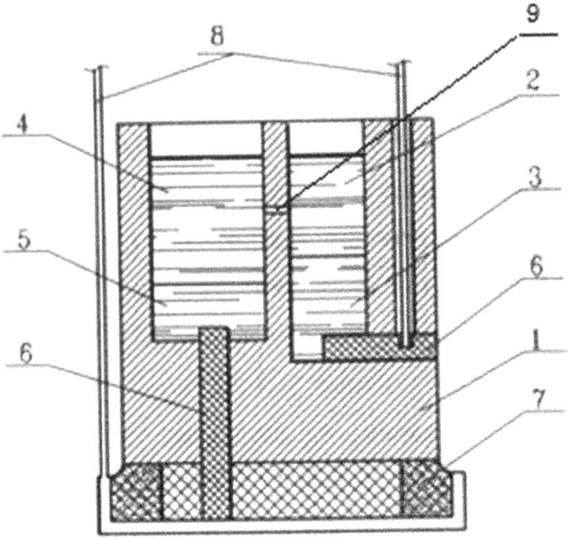

Fig. 2.11 Electrochemical cell for measurements on liquid Cu–Si electrode. Designations: (1) block of boron nitride; (2) melt Na_3AlF_6–Al_2O_3; (3) aluminum; (4) melt Na_3AlF_6–$NaAlSiO_4$; (5) Cu–Si alloy; (6) current leads made of titanium diboride; (7) graphite filling; (8) stainless steel current leads; and (9) a hole of 0.5 mm in diameter

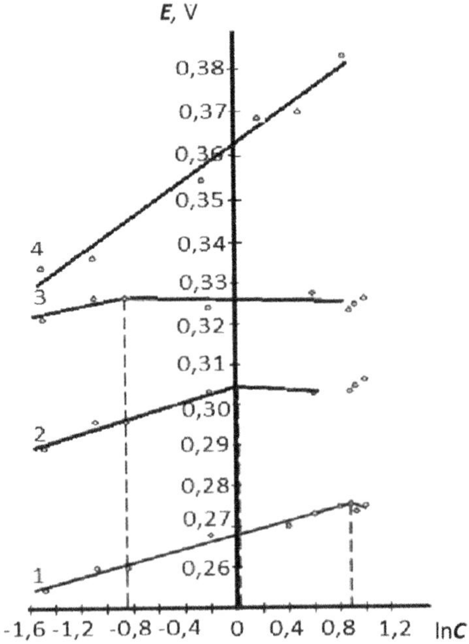

Fig. 2.12 Examples of the dependences of the Si potential in the alloy on the logarithm of the concentration of nepheline at $T = 1300$ K. Designations: (1): 0.10; (2): 0.15; (3): 0.20 at. fraction of Si in the alloy; (4) sodium hexafluorosilicate was used, $C_{Si(Cu)}$: 0.20 at. fraction, $T = 1093$ K

One can see the breaks on the plots, and the greater is the Si content in the alloy, the lower is the SiO_2 content (recalculated from nepheline) in cryolite at the break point.

In our opinion, at the break point on the E–ln C graphs, the surface of the Cu–Si alloy is passivated by a poorly soluble compound $[Si(II)]_{solid}$, which was identified in the Al | Na_3AlF_6–Al_2O_3–SiO_2 system (Figs. 2.5 and 2.6).

The recalculation of the experimental data to pure Si according to the Nernst equation shows that its potential should be 0.42–0.45 V nobler than that for Al, which is close to the results obtained in [41]. The slope of the initial parts of the potentiometric graphs $E = \frac{RT}{nF} \ln C$, where n is the average oxidation number of Si species in cryolite, was within 8–10 mV. Then, $n = 11$–13, which, in accordance with [47], indicates the existence of polynuclear silicon complexes. This agrees with the interpretation of the liquidus line of the Na_3AlF_6–$NaAlSiO_4$ phase diagram, when it was concluded that polyanions exist in the melt in the form of cycles of six alternating $[AlO_4]$ and $[SiO_4]$ tetrahedra (three of each) (see Chap. 1). In other words, the obtained value $n = 11$–13 confirms that the ion-molecular complex participating in the equilibrium electrochemical reaction contains three silicon atoms (see Fig. 1.11).

For comparison, measurements were made in which sodium or potassium hexafluorosilicate M_2SiF_6 (M = Na, K) was the source of silicon. In order to avoid thermal dissociation of the latter, a low-melting cryolite melt (c.r. = 1.5) was used, which made it possible to reduce the temperature of the experiments to $T = 1093$ K. The slopes of the lines are equal to RT/nF. In case of hexafluorosilicate, n turned out to be close to 4 (Fig. 2.12), which indicates different chemical nature of silicon ions present in the melt. If Al_2O_3 was present in the melt, then with the addition of M_2SiF_6, the slope of the graph, as in the case of nepheline, was about $n \sim 12$, and this confirms the tendency of exchange reactions in the reciprocal system Na, Al, Si//O, F to reach the anionic composition corresponding to the stable diagonal cryolite–nepheline (see Chap. 1). It should also be noted that, within the limits of the concentrations of silicon fluoride used, a linear dependence of E–ln C was observed, which means that there was no blocking of the liquid electrode surface. Therefore, it can be assumed that the formation of poorly soluble compounds in an intermediate oxidation state is inherent in electrolytes with a mixed anionic composition, for example, (F, O), (Cl, F), or (Cl, F, O) [7].

To determine the fraction of compounds in the lowest oxidation state Si(II) in the total amount of silicon-containing ions in cryolite, the anodic dissolution [48] of the Cu–Si alloy in the Na_3AlF_6–Al_2O_3 melt was studied. Current density range of 0.005–0.01 A/cm^2 was used, and the amount of electricity passed through the cell was set so that the SiO_2 content in the cryolite after the experiments did not exceed the values corresponding to the inflections on the E–ln C graphs (Fig. 2.12), to escape blocking of the interphase boundary. As it turned out, the average oxidation number of Si passing into the electrolyte at a temperature of $T = 1300$ K is $n = 3.95$–3.98. Consequently, during the electrolysis of the Na_3AlF_6–Al_2O_3–SiO_2 melt, the fraction of intermediate Si(II) compounds is small compared to the content of SiO_2.

Careful analysis revealed that the product of the SiO_2 concentration in cryolite (recalculated from nepheline) and the activity of Si in the alloy, at which an inflection occurred on the E–ln C graphs, were approximately the same (Table 2.1).

Table 2.1 Results of potentiometric measurements, $T = 1300$ K

No.	$\frac{C_{Si(alloy)}}{a_{Si(alloy)}[45.46]}$ (at. fraction)	C_{SiO_2} $\left(\frac{mass\%}{mol\ fract}\right)$	$[C_{SiO_2}]\{a_{Si(Cu)}\}$ ($\times 10^4$)
1	0.10/0.003	1.11/(39.4 × 10^{-3})	1.18
2	0.15/0.009	0.43/(15.2 × 10^{-3})	1.37
3	0.20/0.021	0.17/(6.0 × 10^{-3})	1.26
4	0.25/0.043	0.086/(3.0 × 10^{-3})	1.23
		Average	(1.27 ± 0.25) × 10^{-4}

Evidently, the value

$$\left[C_{SiO_2}\right]\left\{a_{Si(alloy)}\right\} \geq (1.27 \pm 0.25) \times 10^{-4}, \tag{2.6}$$

where the concentration of SiO_2 is in mol fraction and activity of Si in at fraction (and the same applies in the forthcoming parts of the present work), should be considered as a condition for the onset of passivation of the cathode surface by $[Si(II)]_{solid}$ compounds. Based on the most probable [5] intervalence equilibrium:

$$[Si(IV)] + \{Si(alloy)\} \leftrightarrows 2[Si(II)]. \tag{2.7}$$

Hence, this criterion is the equilibrium constant of the process of precipitation dissolution of the Si(II) intermediate. In the chemistry of aqueous solutions for salts with limited solubility, the concept of solubility product is introduced SP = $[M^+][An^-]$, which characterizes the conditions for the formation/dissolution of a precipitate of a given salt [49]. Then, similarly in physical sense, expression (2.6) is the solubility product of the Si(II). Let us call it the passivation criterion $PC_{Si(II)} = [C_{SiO_2}]\{a_{Si(alloy)}\}$.

Since passivation occurs when condition (2.6) is met, then during studies on solid cathodes (i.e., when $a_{Si} = 1$), the $[Si(II)]_{solid}$ deposit will form at a vanishingly low content of SiO_2 in cryolite (~ 10^{-3} mass%).

In accordance with this concept, the effect of the cathode material on the shape of the polarization curves becomes clear (Fig. 2.10). Obviously, the $[Si(II)]_{solid}$ deposit at high temperatures will interact with the electrode surface, and, therefore, depending on the chemical nature of the latter, various effects will be recorded on the *i*–*E* curves. Therefore, the use of solid cathodes to study the reduction of not only silicon, but also other polyvalent metals from a cryolite-oxide melt (as well as from other melts with a mixed anionic composition) makes it possible to obtain information only about the initial stage of the process. The further formation of poorly soluble compounds and their interaction with the cathode material distorts the polarization dependences and

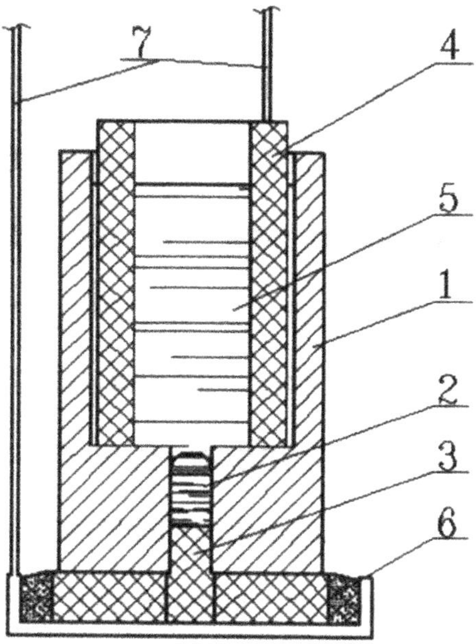

Fig. 2.13 Cell for alternating current measurements on a liquid Cu–Si electrode. Designations: (1) boron nitride crucible; (2) Cu–Si alloy; (3) current lead made of titanium diboride; (4) graphite electrode; (5) melt Na_3AlF_6–$NaAlSiO_4$; (6) graphite filling; (7) stainless steel current leads

cannot yet be quantified. The electrolytic deposition of individual Si (as well as Ge, Ti, Zr, and others) in the form of a metal-salt "rod" proceeds according to the film electrochemical mechanism, the concepts of which, as applied to ionic melts, were developed in [7].

Determination of the kinetic parameters of the Si electrode reaction in the Na_3AlF_6–$NaAlSiO_4$ melt was carried out by the Faraday impedance method in a cell, which is shown in Fig. 2.13.

The Cu–Si alloy was placed in a cylindrical recess 4 mm in diameter, and a graphite tube served as the auxiliary electrode, the working surface area of which was two orders of magnitude greater than that of the indicator electrode. To expand the temperature range, a fusible melt was used (c.r. = 1.7). The measurements were made using the R-5083 automatic bridge. After correcting for the inductance, the components of the Faraday impedance R_f and C_f were found using the Ershler–Randles method [50].

When the content of SiO_2 (recalculated from $NaAlSiO_4$) in the melt did not exceed the value prescribed by criterion (2.6), the frequency dependences of the Faraday impedance $R_f(\omega \cdot C_f)^{-1} = f\left(\omega^{-1/2}\right)$ were straight and parallel: the imaginary part at $\omega \to \infty$ passed through the origin of coordinates and the real part of the impedance cut off on the y-axis a value equal to the charge transfer resistance R_n (Fig. 2.14).

The measurement results were not reproducible if the condition (2.6) was not fulfilled. The nature of the frequency dependence of R_f and $(\omega C_f)^{-1}$ indicates that in the absence of the blocking of the electrode surface, the diffusion and charge transfer have the lowest rates, and therefore, this method does not allow to obtain any information about the conjugated chemical reactions.

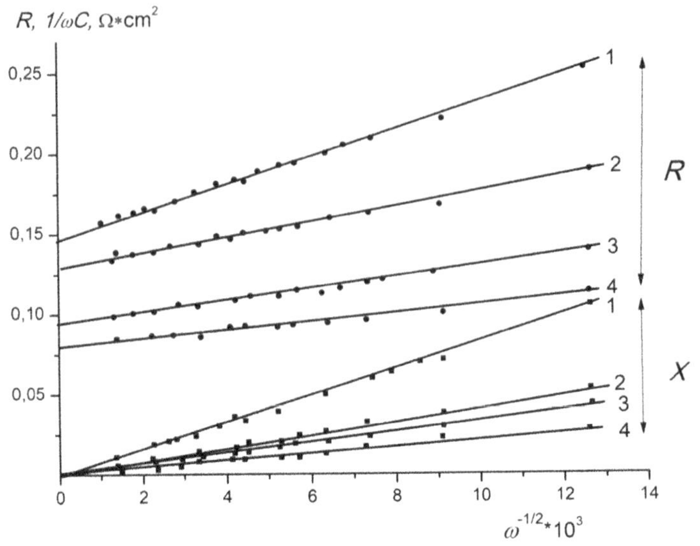

Fig. 2.14 Examples of the frequency dependence of the components of Faraday impedance (*R*-active, *X*-reactive) at $T = 1328$ K. Alloy concentration $C_{Si(Cu)} = 0.1$ at. fractions; content of $NaAlSiO_4$: 1: 0.35 mass%; 2: 0.60 mass%; 3: 1.04 mass%; 4: 1.38 mass%

Kinetic parameters of the electrode reaction were calculated using the equations [51]:

$$R_t = RT/nFi_0 \tag{2.8}$$

$$i_0 = nFk_s(C_{ox})^{1-\alpha}(C_{red})^{\alpha} \tag{2.9}$$

$$E = d(\log i_0)/d(1/T), \tag{2.10}$$

where R_t—charge transfer resistance, Ω cm^2; i_0—exchange current, A/cm^2; E—activation energy of the charge transfer, kJ/mol; k_s—rate constant of the electrochemical reaction, cm/s; α—transfer coefficient, fractions of a unit; C_{ox}—concentration of SiO_2 in cryolite, mol/cm^3; C_{red}—concentration of Si in the alloy, mol/cm^3; other symbols are common.

Since the measurements were performed under conditions when equilibrium (2.7) is established, $n = 4$ is accepted. The results are presented in Table 2.2.

For verification of these results, a selective study of the electrochemical system was carried out by a double-pulse galvanostatic method. The cell for measurements did not fundamentally differ from that shown in Fig. 2.11, with the only difference that the volume of the indicator electrode compartment was increased, and the surface of the liquid metal was reduced by placing the latter in a channel with a diameter of 3 mm. A graphite auxiliary electrode was introduced into the same compartment, the area

Table 2.2 Kinetic parameters of the electrode reaction of silicon in the cryolite-oxide melt according to the data of impedance measurements

$C_{Si(alloy)}$ (at. fract)	$C_{NaAlSiO_4}$ ($\frac{mass\%}{mol/cm^3}$)	T (K)	i_0 (A/cm^2)	$k_s \times 10^3$ (cm/s)	α	E_{i_0} (kJ/mol)
0.10	$\frac{0.35}{5.03\times10^{-5}}$	1328	0.21	2.81	0.63	
	$\frac{0.69}{10.02\times10^{-5}}$		0.25			
	$\frac{1.04}{14.98\times10^{-5}}$		0.31			
	$\frac{1.38}{19.9\times10^{-5}}$		0.34			
0.15	$\frac{0.35}{5.01\times10^{-5}}$	1213	0.45	2.55	0.60	
	$\frac{0.69}{9.75\times10^{-5}}$	1213	0.52			59.4
		1263	0.65			
		1303	0.81			
		1338	0.89			
	$\frac{1.04}{14.93\times10^{-5}}$	1213	0.74			

of which was 80 times larger than that for the alloy under study. The measurements were carried out according to the scheme [51], which included a G5-30 two-channel rectangular pulse generator, a C8-13 storage oscilloscope, and a bridge circuit. After selecting the ratio of pulse amplitudes, at which a horizontal platform is observed on the η–t curve, the duration of the first pulse t_1 was varied and a graph $\eta - \sqrt{t_1}$ was plotted. By extrapolating this graph to zero time, we obtain the value of the transition overvoltage η. The exchange current i_0 was calculated using the formula:

$$i_0 = RTi_2/nF\eta \tag{2.11}$$

where i_2—current density of the second pulse, A/cm^2.

The transfer coefficient, rate constant, and activation energy were found from Eqs. (2.9) and (2.10). These data are given in Table 2.3.

Comparison of the results obtained by the Faraday impedance and two-pulse galvanostatic methods shows their fairly good correlation.

2.6 Mechanism of the Alloy Formation in the Aluminum Electrolyzer

The reduction of silicon compounds with liquid aluminum from the Na_3AlF_6–Al_2O_3–SiO_2 melt is, in its physico-chemical nature, a "cementation" reaction, that is, anodic dissolution of Al and cathodic discharge of Si(IV) occur simultaneously. Let us consider how this process occurs during the cathodic polarization of aluminum, as it is in an industrial electrolyzer. The information about it can be obtained using

Table 2.3 Kinetic parameters of the electrode reaction of silicon in the cryolite-oxide melt according to the data of the double-pulse galvanostatic method for the alloy $C_{Si(Cu)} = 0.15$ at. parts

$C_{NaAlSiO_4}$ $\left(\frac{mass\%}{mol/cm^2}\right)$	T (K)	i_0 (A/cm^2)	$k_s \times 10^3$ (cm/s)	α	E_{i_0} (kJ/mol)
$\frac{0.24}{3.41\times10^{-5}}$	1213	0.31			
$\frac{0.40}{5.70\times10^{-5}}$	1213	0.36	2.52	0.66	
$\frac{0.64}{9.12\times10^{-5}}$	1213	0.43			
$\frac{0.60}{11.40\times10^{-5}}$	1213	0.46			63.8
	1243	0.53			
	1273	0.61			
	1303	0.70			

diagrams of exchange currents i_0 and current potential i–E (Figs. 2.15 and 2.16, respectively).

The exchange current of aluminum in the Na_3AlF_6–Al_2O_3 melt is $i_{0(Al)}$ ~ 12 A/cm^2 [52] (Fig. 2.15a). In the presence of SiO_2, the aluminum potential $E_{Al}{}^0$ is shifted to a stationary value E_{st} (Fig. 2.2) due to the occurrence of simultaneous redox reactions $i_{an(Al)} = i_{d(Si)}$ (Fig. 2.16).

In an industrial electrolyzer, it is possible to obtain an Al (~ 8 mass%)–Si alloy, for which the silicon activity is $a_{Si(Al)}$ ~ 0.011 [30–32]. For the latter value, according to Tables 2.2 and 2.3, the exchange current density $i_{0(Si)}$ is equal to 0.6–0.7 A/cm^2, and, at the content of SiO_2 in cryolite ~ 0.5 mass%, it will be an order of magnitude higher than the limiting current Si ($i_{d(Si)}$ ~ 0.05 A/cm^2). It means that its reduction will be accompanied by concentration polarization, that is, it will proceed in the diffusion kinetics regime [53]. The observed current values are the difference between the partial current values (Fig. 2.15b):

$$i_{an(Al)} = i_{d(Si)} = \left|i_{an(Al)} - i_{cat(Al)}\right| = \left|i_{cat(Si)} - i_{an(Si)}\right|. \quad (2.12)$$

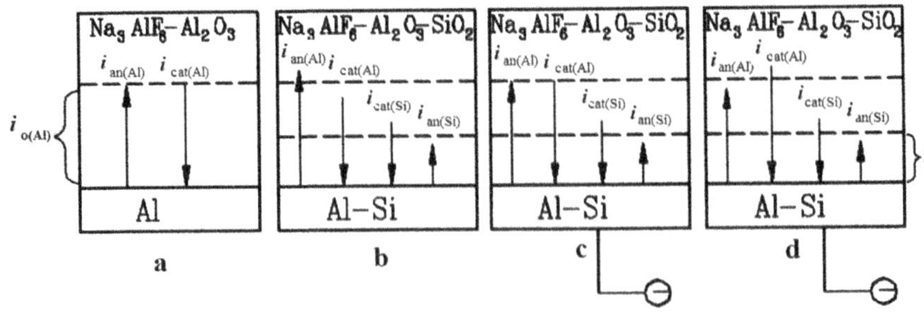

Fig. 2.15 Current diagrams of the alloy formation. See explanations in the text

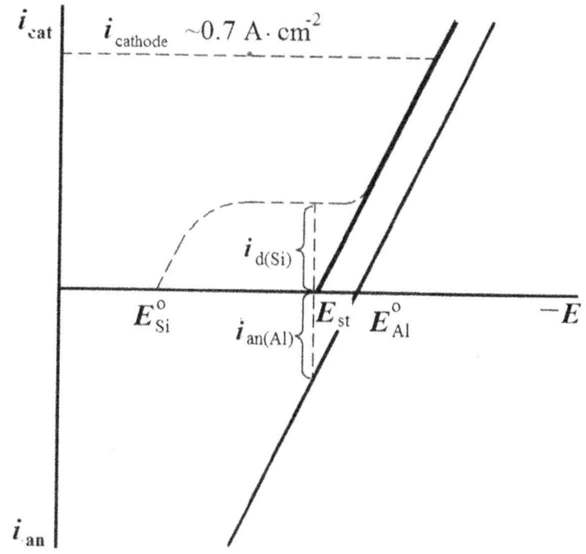

Fig. 2.16 Mechanism of alloy formation on an aluminum cathode presented as a current–potential (i–E) diagram. The bold line indicates the total hypothetical polarization curve

Under cathodic polarization of the liquid electrode, the exchange current of aluminum $i_{an(Al)}$ decreases, and the equilibrium potential $E_{Al}{}^0$ will be reached (as for diffusion-controlled processes) at an external current density equal to the limiting current of silicon $i_{cathode} = i_{dSi}$ [54] (Figs. 2.15c and 2.16). In industrial electrolyzers, the current density reaches $i_{cathode} > 0.7$ A/cm^2, which means that aluminum is cathodically protected (Fig. 2.15d), and the alloying process obeys the laws of joint ion discharge and can be represented by a hypothetical total polarization curve (Fig. 2.16).

2.7 Conclusions

The aluminothermic reduction of silicon compounds from the Na_3AlF_6–Al_2O_3–SiO_2 melt proceeds according to the electrochemical mechanism, i.e., it is a combination of reactions of anodic dissolution of Al and cathodic discharge of Si(IV). With the cathodic polarization of aluminum, as in an industrial electrolyzer, its dissolution is suppressed and the process of alloy formation will obey the laws of the joint discharge of ions. The actual reduction of Si(IV) proceeds stepwise, and its intermediate has a limited solubility, above which a passivating deposit $[Si(II)]_{solid}$ will form at the interface. The condition for the precipitation of the latter is the excess of its passivation criterion $PC_{Si(II)} = [C_{SiO_2}][a_{Si(Cu)}] = (1.27 \pm 0.25) \times 10^{-4}$. It is obvious that when developing a technology for producing Al–Si alloys, it is necessary to find such means of loading SiO_2 into an electrolytic cell which eliminate this negative phenomenon.

Given that the reduction of other polyvalent metals from a cryolite-oxide melt also results in the formation of a precipitate of their poorly soluble compounds in an intermediate oxidation state, the regularities established for silicon with the quantitative characteristics inherent in each of these metals will obviously be valid for them as well.

References

1. Sukhotin AM (1981) Spravochnik po elektrokhimii (Handbook on electrochemistry). Khimiya, Leningrad, 488 p (in Russian)
2. Efimov EA, Erusalimchik IG (1963) Elektrokhimiya germaniya i kremniya (Electrochemistry of germanium and silicon). Goskhimizdat, Moscow, 180 p (in Russian)
3. Minet A (1891) Comp Rend 112:1215–1218
4. Esin OA (1970) Fizicheskaya khimiya rasplavlennykh shlakov (Physical chemistry of molten slags). Naukova Dumka, Kiev, pp 5–34 (in Russian)
5. Novokshonov NI, Nikitin YuP, Novoladskiy VP, Vlasov NN (1974) Elektrokhimiya i rasplavy (Electrochemistry and melts). Nauka, Moscow, pp 79–82 (in Russian)
6. Delimarskiy YuK (1980) Khimiya ionnykh rasplavov (Chemistry of ionic melts). Naukova Dumka, Kiev, 328 p (in Russian)
7. Andriiko AA, Andriiko YuA, Nauer GE (2013) Many electron electrochemical process, vol XIX. Springer, 167 p
8. Samsonov HV (1965) Fiziko-khimicheskiye svoistva elementov (Physico-chemical properties of elements). Naukova Dumka, Kiev, 808 p (in Russian)
9. Geld PV, Esin OA (1957) Protsessy vysokotemperaturnogo vosstanovleniya (High temperature reduction processes). Metallurgizdat, Sverdlovsk, 646 p (in Russian)
10. Kozhevnikov GN, Vodopianov AG (1977) Nizshye okisly kremniya i aliuminiya v elektrometallurgii (Low valency oxides of silicon and aluminum in electrometallurgy). Nauka, Moscow, 145 p (in Russian)
11. Rusakov LN, Dubrovin AS (1963) Proc USSR Acad Sci 149:107–110 (in Russian)
12. Rusakov LN, Dubrovin AS, Lyakishev NP (1972) Commun USSR Acad Sci Met 2:31–36 (in Russian)
13. Belyaev AI (1947) Fiziko-khimicheskiye protsessy pri electrolyze aluminiya (Physico-chemical processes in aluminum electrolysis). Metallurgizdat, Moscow, 183 p (in Russian)
14. Monier R, Barakat D (1957) Helv Chim Acta 40:2041–2045
15. Grijotheim K, Matiosovsky K, Fellner P (1972) Rev Rom Chim 17:819–829
16. Frazer EJ, Welsh DJ (1977) Electrochim Acta 22:1179–1182
17. Brönsted J, Kane N (1931) J Am Chem Soc 53:3624–3644
18. Frumkin AN (1932) Z Phys Chem A 160(2):116–118
19. Hammet L, Lorch A (1932) J Am Chem Soc 54:2128–2129
20. Kaesche H (1979) Die Korrosion der Metalle. Springer-Verlag, Berlin, Heidelberg, New York, 400 p
21. Antropov LI, Donchenko MI (1973) Itogi nauki i tekhniki. Korroziya i zaschita ot korrozii (Achievements in science and technology. Corrosion and corrosion protection), vol 2. VINITI, Moscow, pp 113–170 (in Russian)
22. Delimarskii YuK, Pruttskov DV, Andriiko AA, Chernov RV (1983) Ukr Chem J 49:739–742 (in Russian)
23. Pruttskov DV, Andriiko AA, Chernov RV, Delimarskii YuK, Khvalin AP (1983) Ukr Chem J 49:845–849 (in Russian)
24. Pruttskov DV, Pirozhkova VP, Khvalin AP, Chernov RV, Delimarskii YuK (1983) Ukr Chem J 49:1027–1030 (in Russian)

25. Pruttskov DV, Pirozhkova VP, Chernov RV, Khvalin AP (1984) Ukr Chem J 50:1082–1085 (in Russian)
26. Pruttskov DV, Andriiko AA, Delimarskii YuK, Chernov RV (1985) Ukr Chem J 51:826–830 (in Russian)
27. Pruttskov DV (1986) Ionnye rasplavy i tverdye electrolity (Ionic melts and solid electrolytes), vol 1. Naukova Dumka, Kiev, pp 70–77 (in Russian)
28. Pruttskov DV, Andriiko AA, Chernov RV (1987) Tsvetnye Met (Non-ferrous Met) 2:39–43 (in Russian)
29. Pruttskov DV, Krivoruchko NP, Olesov YuG (1994) Ukr Chem J 60:433–439 (in Russian)
30. Berton O, Petot-Ervas G, Petot C, Desre P (1969) Comp Rend Ser C 268:1939–1942
31. Batalin GI, Beloborodova EA, Stukalo VA (1971) Commun USSR Acad Sci Met 2:69–74 (in Russian)
32. Schaefer SC (1974) Rept Invest Bur Mines US Dep Inter 7895:1–15
33. Stroganov GB, Rotenberg VA, Gershman GB (1977) Splavy aliuminiya s kremniyem (Alloys of aluminum with silicon). Metallurgy, Moscow, 272 p (in Russian)
34. Gorji YM, Soltanieh M, Habibolahzaden A (2007) Can Metall Q 46:385–390
35. Winchell AN, Winchell H (1964) The microscopical characters of artificial inorganic solid substances. Optical properties of artificial minerals. Academic Press, New York, 439 p
36. Milov AI (1980) Kompleksnaya pererabotka rud tsvetnykh metallov (Complex processing the ores of non-ferrous metals). Science, Alma-Ata, pp 98–107 (in Russian)
37. Belyaev AI (1944) Metallurgiya legkikh metallov (Metallurgy of light metals). Metalurgistdat, Moscow, 843 p (in Russian)
38. Nerubashchenko VV, Krymov AP, Voleinik VV (1977) Tsvetnye Met (Non-ferrous Met) 7:29–31 (in Russian)
39. Nerubashchenko VV, Voleinik VV, Krymov AP (1978) Tsvetnye Met (Non-ferrous Met) 3:36 (in Russian)
40. Zakharov MS, Bakanov VI, Pnev VV (1978) Khronopotentsiometriya (Chronopotentiometry). Khimiya, Moscow, 200 p (in Russian)
41. Khushkhov KhB, Malyshev VV, Gasviani SG, Shapoval VI, Gasviani NA (1991) Ukr Chem J 57:1097–1100 (in Russian)
42. Grjotheim K, Brdicka P, Silny A, Matiasovsky K, Stubergh J (1991) Can Metall Q 30:107–111
43. Sokhanvaran S, Barati M (2014) J Electrochem Soc 161(1):E6–E11
44. Liu A, Shi Z, Hu X, Gao B, Wang Z (2017) J Electrochem Soc 164(2):H126–H133
45. Chart TG (1973) High Temp High Press 5:241–252
46. Beloborodova EI, Zinkevich TN, Gab AI, Pruttskov DV, Bondarenko GN (1994) Protsessy litia (Casting processes), vol 2, pp 55–65 (in Russian)
47. Kravtsov VI (1985) Ravnovesiya i kinetika elektrodnykh reaktsiy kompleksov metallov (Equilibrium and kinetics of electrode reactions with metal complexes). Khimiya, Leningrad, 208 p (in Russian)
48. Delimarskiy YuK, Zarubitskiy OG (1971) Proc USSR Acad Sci Ser B 8:709–710 (in Russian)
49. Dickerson RE, Gray HB, Haight GP (1979) Chemical principles, 3rd edn. The Benjamin/Cummings Publishing Company, Inc., Menlo Park, CA. ISBN 0805323988. https://resolver.caltech.edu/CaltechBOOK:1979.001
50. Grafov BM, Ukshe EA (1973) Elektrokhimicheskiye tsepi peremennogo toka (Electrochemical circuits of alternating current). Nauka, Moscow, 128 p (in Russian)
51. Damaskin BB (1965) Printsipy sovremennykh metodov izucheniya elektrokhimicheskikh reaktsiy (Principles of the modern methods for studies of electrochemical reactions). MGU, Moscow, 104 p (in Russian)
52. Thonstad J, Rolseth S (1978) Electrochim Acta 23:223–241
53. Gorodysskiy AV (1988) Voltamperometriya, Kinetika statsionarnogo elektroliza (Voltamperometry: kinetics of stationary electrolysis). Naukova Dumka, Kiev, 176 p (in Russian)
54. Antropov LI (1984) Theoretical electrochemistry. Vysshayashkola, Moscow (in Russian) (English edition: 1972, Mir Publishers, Moscow)

Chapter 3
Current Yield

3.1 Introduction

The actual productivity of an aluminum electrolyzer is always less than the theoretical one calculated according to Faraday's law. As indicated in the fundamental publications [1–3], the main reasons for the decrease in current efficiency are:

- evaporation of the products of the interaction of aluminum and cryolite;
- dissolution of aluminum in cryolite and its subsequent oxidation.

It is obvious that when obtaining Al–Si alloys, these factors will also be of decisive importance.

All studies of vapor pressure and its composition were performed by indirect tensiometric methods in combination with gravimetry and chemical analysis of the condensate [1–3]. In the cryolite (c.r. = 3)/Al system, abnormally high vapor pressure develops ($T = 1300$ K, $p = 10.5$ kPa [2]), which is not common for pure cryolite ($T = 1300$ K, $p = 0.5$ kPa [2]).

This is due to the fact that, in addition to tetrafluoroaluminate $NaAlF_4$ (the product of cryolite evaporation), Na, AlF subfluoride, and, possibly, the hypothetical compound Na_2F appear in the gas phase. They are the result of the redox reactions

$$\mathrm{Al} + 6\mathrm{NaF} = \mathrm{Na_3AlF_6} + 3\mathrm{Na}_{(\mathrm{vapor})} \tag{3.1}$$

$$2\mathrm{Al} + \mathrm{Na_3AlF_6} = 3\mathrm{NaF} + 3\mathrm{AlF}_{(\mathrm{vapor})}. \tag{3.2}$$

While the existence of Na in a vapor is proved by measuring the absorption of ultraviolet radiation [4], the presence of AlF is related to the detection of pure aluminum particles in sublimes formed because of its disproportionation [5]. As for Na_2F, no experimental data on its existence have been found. Unambiguous information about the vapor composition can be obtained by the Knudsen effusion method combined with mass spectra [6].

D. Pruttskov et al., *Electrolytic Production of Al–Si Alloys*,
Monographs in Electrochemistry, https://doi.org/10.1007/978-3-031-29249-1_3

The solubility of aluminum in cryolite was considered in [7]. Dissolved aluminum is supposed to be the total content of univalent aluminum and sodium in the fluoride melt, formed by similar redox reactions

$$2Al + Na_3AlF_6 \leftrightarrows 3NaAlF_{2(solution)} \tag{3.3}$$

$$Al + 6NaF \leftrightarrows Na_3AlF_6 + 3Na_{(solution)}. \tag{3.4}$$

Its solubility in cryolite (c.r. = 3) at $T = 1273$ K, recalculated on aluminum is, in mass%: $C_{Al} = 0.082$ and, by components, $C_{Na} = 0.020$ and $C_{NaAlF_2} = 0.062$. Similar values are reported in [8]: 0.036 and 0.063, respectively. With a decrease in the activity of the metal due the formation of alloy with a less active element, its solubility in the ionic melt decreases, and, accordingly, the rate of oxidation [9] also decreases. This phenomenon has been confirmed for Al–Cu alloy in contact with cryolite [2]. Silicon, like copper, is a more noble metal than aluminum, and therefore its effect on the solubility and, in turn, on the loss of the Al–Si alloy should be similar, which requires experimental verification.

This chapter summarizes the results of the authors' studies on these problems [10–13].

3.2 Loss of Metal in the Al–Si/Na_3AlF_6 System

Information on weight loss in the Na_3AlF_6/Al–Si system was obtained by weighing a TiB_2 crucible with ingredients using an E-2D1 mechanoelectric converter. Examples of thermogravigrams are shown in Fig. 3.1.

The evaporation rate was determined from the slope of the tangent to the initial section of the curves. The loss of the alloy was found from the change in the mass of its initial sample over 3 h of exposure. As follows from Fig. 3.2, both the evaporation rate of the metal-salt mixture and the loss of the alloy, depending on its composition, pass through a minimum.

Losses are presented in relative fractions, where those for aluminum are taken as one. More detailed studies were required to interpret the dependences obtained.

3.3 Studies of Vapor Composition by the Method of Mass Spectroscopy

The effusion method followed by the study of the mass spectrum allows direct determination of the vapor composition and pressure, but it suffers from a disadvantage. The point is that the maximum pressure developed inside the Knudsen cell at which the molecular outflow of vapor through the effusion hole still occurs (i.e., collisions

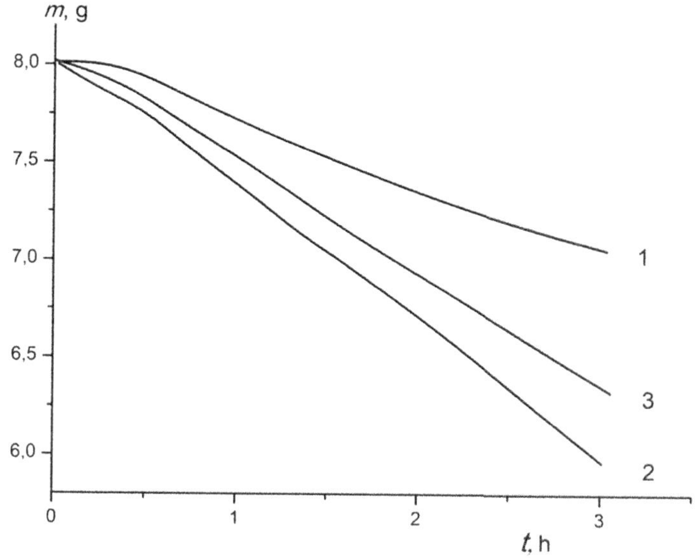

Fig. 3.1 Examples of thermogravigrams of the Na_3AlF_6/Al–Si system. Designations: (1) cryolite; (2) cryolite/aluminum; (3) cryolite/alloy Al (15 mass%)–Si. $T = 1303$ K; evaporation area 3.8 cm^2; initial weight 8 g; the ratio of reagents is 1:1

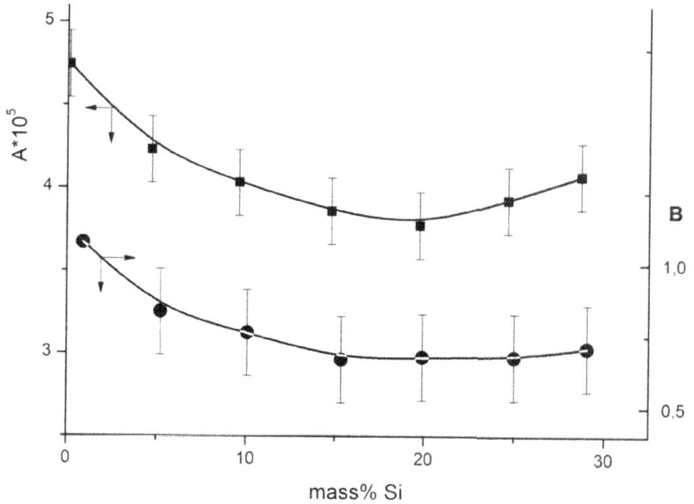

Fig. 3.2 Dependence of the evaporation rate of the Na_3AlF_6/Al–Si system and the relative losses of the alloy on its concentration. Designations: A (left)—evaporation rate, g/cm^2 s; B (right)—fraction of alloy losses relative to the losses of pure aluminum

between molecules in the channel can be neglected) is < 10 Pa [14]. For a cryolite melt, and even more so in the presence of aluminum, the vapor pressure exceeds this value by several orders of magnitude. Therefore, measurements were made at a temperature of 950 K, when fluorides were solid and only aluminum (and Al–Si alloys) was liquid, because only under these conditions are it possible to use this method.

The experiments were carried out on MX-1303 and MI-1201 devices equipped with a high-temperature ion source. The scheme of the high-temperature mass spectral experiment and the design of the high-temperature ion source are shown in Figs. 3.3 and 3.4.

Evaporation was carried out using effusion cells: double platinum and single-section graphite and titanium cells coated with TiB_2. To determine the molecular precursors of the ion current, ionization efficiency curves (IECs) were plotted; i.e., the dependence of the ion current on the energy of ionizing electrons was found, which was recorded on a two-coordinate recorder LKD-4-003 and analyzed according to the method [6]. The time dependence of the ion current was recorded on a KSP-4 potentiometer.

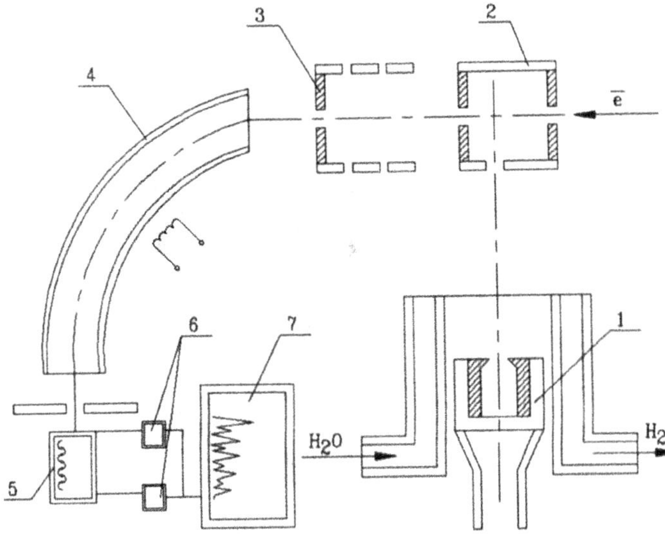

Fig. 3.3 Scheme of a high-temperature mass spectral experiment. Designations: (1) effusion cell; (2) ionization chamber; (3) system of pulling lenses; (4) mass analyzer; (5) ion current collector; (6) amplifiers; (7) fixing device

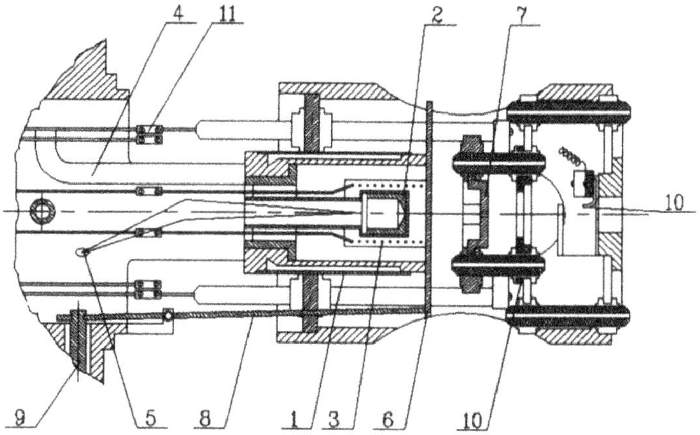

Fig. 3.4 Design of a high-temperature ion source. Designations: (1) water-cooled block; (2) effusion cell; (3) heater; (4) water supply; (5) thermocouple; (6) movable diaphragm; (7) ionization chamber; (8) different-shoulder lever for moving the diaphragm; (9) bellows mechanism; (10) lens system; (11) current leads to power consumers

3.4 System Al/Na_3AlF_6

Mass spectral studies of the gas phase over MF/AlF_3 mixtures showed that at the ratio $MF/AlF_3 < 3$ in a vapor, the main components are $MAlF_4$ tetrafluoroaluminate and a small amount of its dimer $(MAlF_4)_2$, which, upon electron impact, decompose into positive M^+ ions, $MAlF_3^+$, $M_2AlF_4^+$ and AlF_2^+. At $MF/AlF_3 > 3$ in the vapor, along with $MAlF_4$, alkali metal fluoride MF and its dimer $(MF)_2$ appear, which are formed upon ionization of M^+, MF^+ and M_2F^+ [6].

To obtain comparative data, systems with sodium, potassium, rubidium and cesium (M) cryolites ($MF–AlF_3/Al$) were studied. The mass spectra of the gas phase, depending on the MF/AlF_3 ratio, consisted of M^+, AlF^+, MF^+, M_2F^+, $MAlF_3^+$, $M_2AlF_4^+$ and AlF_2^+ ions. Comparison of them with the mass spectra of saturated vapor of purely salt systems MF/AlF_3 [6] showed that, when aluminum was added, a new AlF^+ line appeared and the intensity of the M^+ ion current increased noticeably (Table 3.1).

Table 3.1 Mass spectra of the saturated vapor of the $NaF–AlF_3$ (c.r. = 1.5)/Al system in a graphite cell

Composition	Ions, $I_{rel.}$			
	Na^+	AlF^+	AlF_2^+	$NaAlF_3^+$
$NaF–AlF_3$	100	–	3.9 ± 0.2	1.2 ± 0.1
$NaF–AlF_3/Al$	100	45.2 ± 2.6	0.63 ± 0.08	0.22 ± 0.04

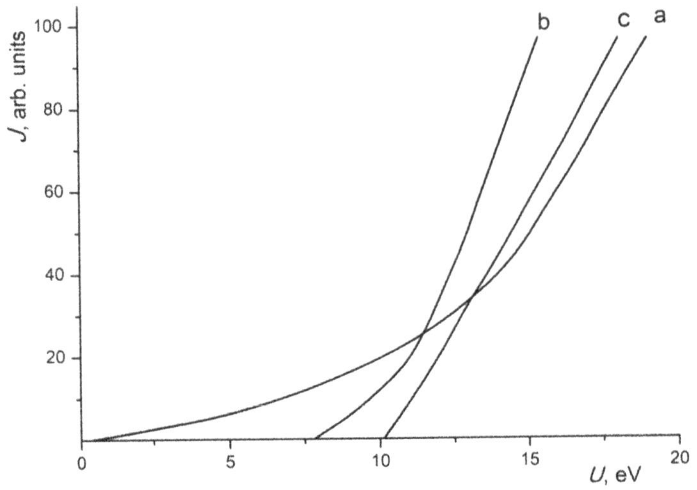

Fig. 3.5 Ionization efficiency curves for ion currents. Designations: (a) sodium; (b) aluminum subfluoride; (c) particles with $m/e = 65$

The rest of the recorded currents and the ratios between them characterize the fragmentation products of the components of the salt phase in the absence of aluminum.

The survey of the IEC showed that the values of the ion current of alkali metals M^+ exhibit different character than those for particles with $m/e = 46$ and 65 (Fig. 3.5).

In all systems at $MF/AlF_3 < 3$, after the initial rise in current at 4–5 eV, changes of the slope were observed on the curves at 13–14 eV. This shape of IEC allows us to assume that the M^+ ion is formed from two precursors, namely M atoms and $MAlF_4$ molecules. The appearances of AP ($M^+/MAlF_4$) agree with the literature data. At $MF/AlF_3 > 3$, the M^+ ion should be obtained from three sources MF, $MAlF_4$ [6] and M. However, due to the significant predominance of the metal in saturated vapor, the inflections corresponding to the decomposition of MF and $MAlF_4$ were not clearly fixed on the IEC of the ion current M^+.

The shape of IEC for an ion with $m/e = 46$ in all MF-AlF_3/Al systems (Fig. 3.5) allowed us to conclude that it comes from only one molecule, namely from AlF. This is also evidenced by the obtained value of the appearance potential (AP) (AlF^+/AlF), equal to 9.8 ± 0.2 eV, which coincides with the reference value 9.7 ± 0.5 eV [15] (Table 3.2).

It was of interest to study the current for an ion with $m/e = 65$ in the NaF–AlF_3/Al system, for which Na_2F can be a precursor. The IEC survey was performed for the "alkaline" electrolyte composition $NaF/AlF_3 > 3$, where its appearance should be more probable (Fig. 3.5). The current corresponds to one ionization process, and the value of the potential (10.7 ± 0.2 eV) coincides with the value of the appearance potential AP (Na_2F^+/Na_2F_2) for the $(NaF)_2$ dimer, which was measured in a special experiment. In this experiment, NaF was evaporated, the dimer of which is

Table 3.2 Ionization potentials IP (M^+/M) and appearance potentials AP ($M^+/MAlF_4$) for $MF–AlF_3/Al$ systems

Metal	IP (M^+/M) (eV)		AP ($M^+/MAlF_4$) (eV)	
	Experiment	Reference [15]	Experiment	Reference [16]
Na	5.2 ± 0.2	5.14	13.7 ± 0.3	13.50
K	4.5 ± 0.3	4.34	13.0 ± 0.3	13.15
Rb	4.3 ± 0.5	4.18	12.7 ± 0.2	12.94
Cs	4.0 ± 0.5	3.89	12.9 ± 0.2	12.93

present in a noticeable amount in the gas phase [6]. Based on physical considerations, one can also expect that the Na_2F^+ ion current from Na_2F would appear at a lower ionization energy than during the decomposition of $(NaF)_2$, i.e., IP (Na_2F^+/Na_2F) < AP (Na_2F^+/Na_2F_2) [6] and an inflection should have appeared on the IEC (as for the Na^+ current with two precursors, Fig. 3.5), which was not actually observed. Hence it follows that, there are no Na_2F molecules in the saturated vapor of the $NaF–AlF_3/Al$ system, at least within the sensitivity of the mass spectral method. Subfluorides of other alkali metals for the corresponding systems were also not found.

Calculations of the partial pressures of the vapor components over the Al/Na_3AlF_6 system were performed using the formula [6]

$$p_{\mathrm{j}} = K I_{\mathrm{j}} T / \sigma_{\mathrm{j}}, \tag{3.5}$$

where K is the sensitivity constant of the device; I_{j} is the total current of all ions formed during the ionization of j-type molecules; T is the absolute temperature; σ_{j} is the total ionization cross section of j-type molecules.

The total ionization cross sections were calculated according to the additivity rule [6] and they amounted to: $\sigma_{\mathrm{Na}} = 4.02 \times 10^{16}$ cm^2, $\sigma_{\mathrm{AlF}} = 7.21 \times 10^{16}$ cm^2, $\sigma_{\mathrm{NaAlF_4}} = 8.05 \times 10^{16}$ cm^2.

The quantitative interpretation was made as follows. The current $I_{\mathrm{Na^+}}$ is formed during the ionization of atomic Na and $NaAlF_4$ molecules (Fig. 3.5)

$$I_{\mathrm{Na^+}} = I_{\mathrm{Na^+/Na}} + I_{\mathrm{Na^+/NaAlF_4}}. \tag{3.6}$$

Separation of $I_{\mathrm{Na^+}}$ into components was carried out by introducing the mass spectrum coefficient for the ion having one precursor, which is $NaAlF_3^+$, the product of dissociative ionization of $NaAlF_4$ [6]. Using the mass spectrum coefficient

$$\alpha = \frac{I_{\mathrm{Na^+/NaAlF_4}}}{I_{\mathrm{NaAlF_3^+/NaAlF_4}}} \tag{3.7}$$

and taking into account that the AlF^+ ion has one precursor, and the $NaAlF_4$ molecule upon electron impact forms Na^+, AlF_2^+ and $NaAlF_3^+$ fragments [6], we obtain the following system of equations:

$$I_{Na^+/Na} = I_{Na^+} - \alpha \cdot I_{NaAlF_3^+/NaAlF_4} \quad (3.8)$$

$$I_{AlF} = I_{AlF^+} \quad (3.9)$$

$$I_{NaAlF_4} = (1 + \alpha) \cdot I_{NaAlF_3^+/NaAlF_4} + I_{AlF_2/NaAlF_4}. \quad (3.10)$$

The determination of α was carried out in a special experiment with a mixture of Na_3AlF_6–$Na_5Al_3F_{14}$. In the vapor above this mixture, there are predominantly $NaAlF_4$ and a much smaller amount of the dimer $(NaAlF_4)_2$. Its content decreases sharply at the temperatures of our experiments [6]. Therefore, we neglect the current $I_{Na^+/(NaAlF_4)_2}$, because $I_{Na^+/NaAlF_4} \gg I_{Na^+/(NaAlF_4)_2}$. That is

$$I_{Na^+} = I_{Na^+/NaAlF_4} + I_{Na^+/(NaAlF_4)_2} \approx I_{Na^+/NaAlF_4} \quad (3.11)$$

Based on the measured I_{Na^+} and $I_{NaAlF_3^+}$, the mass spectrum coefficient was calculated, which was $\alpha = 87 \pm 3$. Table 3.3 shows the absolute values of the intensities of the main lines of the mass spectrum of the saturated vapor of the Al/Na_3AlF_6 system, and Table 3.4 shows the calculated total values of the current intensity of ions formed during the decomposition of molecular precursors.

To calculate partial pressures, it is necessary to know the sensitivity constant K. For a two-section platinum cell, it was determined by the method of standard [6] and amounted to $(1.65 \pm 0.10) \times 10^{-25}$ atm cm^2/K V. Since the method of standard is not applicable for finding K for a single-section graphite cell, the following technique was used [6]. We assume that equilibrium is established in the system under study

$$3Na_3AlF_6 + 2AlF \leftrightarrows 4Na + 5NaAlF_4 \quad (3.12)$$

Table 3.3 Absolute values of the intensities of the main lines of the mass spectra of the saturated vapor of the Al/Na_3AlF_6 system

Cell material	Absolute intensities I (V)			
	Na^+	AlF^+	AlF_2^+	$NaAlF_3^+$
Platinum	74 ± 2	13.1 ± 0.3	0.52 ± 0.01	0.64 ± 0.01
Graphite	24.7 ± 0.4	10.8 ± 0.2	0.19 ± 0.005	0.21 ± 0.005
TiB_2	18.0 ± 0.4	9.2 ± 0.1	0.09 ± 0.002	0.13 ± 0.002

Table 3.4 Calculated total current intensities of ions formed during the decomposition of molecular precursors

Cell material	Total intensities I (V)		
	Na	AlF	$NaAlF_4$
Platinum	18.3 ± 4.8	13.1 ± 0.3	55.7 ± 2.8
Graphite	6.4 ± 1.5	10.8 ± 0.2	18.3 ± 1.1
TiB_2	6.7 ± 0.7	9.2 ± 0.1	11.5 ± 0.5

with equilibrium constant

$$K_{p(3.12)} = P_{Na}{}^4 \cdot P_{NaAlF_4}{}^5 / P_{AlF}{}^2 \ldots . \tag{3.13}$$

Going from partial pressures to ionic current, using Eq. (3.5), we found an expression for the sensitivity constant:

$$K = \frac{1}{T} \sqrt[7]{\frac{\sigma_{Na}{}^4 \cdot \sigma_{NaAlF_4}{}^5 \cdot I_{AlF}{}^2}{\sigma_{AlF}{}^2 \cdot I_{Na}{}^4 \cdot I_{NaAlF_4}{}^5}} \cdot K_{p(3.12)}. \tag{3.14}$$

$K_{p(3.12)}$ was calculated using thermodynamic data [17, 18] (for Na_3AlF_6) and [6] (for $NaAlF_4$). For the graphite cell, $K = (6.2 \pm 1.1) \times 10^{-25}$ atm cm^2/K V.

For a single-section cell coated with TiB_2, the value of K was found by the method of complete isothermal evaporation [6]. From the Knudsen chamber, ingredients with a known mass were evaporated and the intensities of the mass spectrum lines were recorded over time. The sensitivity coefficient of the device was determined by the Hertz–Knudsen equation, in which the pressure is replaced by the ion current:

$$g_j = S(2\pi MRT)^{-1/2} \cdot \int_0^T p\mathrm{d}t = \frac{K}{\sigma_j} \cdot S(2\pi MRT)^{-1/2} \cdot T^{1/2} \int_0^T I_j \mathrm{d}t, \tag{3.15}$$

where g_j is the number of moles of the evaporated substance; S—effective area of the effusion hole; M—molar weight; the other designations are common.

Figure 3.6 shows the intensities of the ion current $I_j = f(\tau)$ of the evaporation products of the Al/Na_3AlF_6 system with an excess of metal.

The integrals were calculated graphically. The symbatic course of the curves indicates the congruence of evaporation, and this process can be described by the overall reaction:

$$Na_3AlF_6 + 2Al \leftrightarrows NaAlF_4 + 2AlF + 2Na. \tag{3.16}$$

Then, from the material balance of this reaction

$$q_{Na_3AlF_6} + 2q_{Al} = q_{NaAlF_4} + 2q_{AlF} + 2q_{Na} \tag{3.17}$$

And, knowing the quantity of cryolite moles, from Eq. (3.15) we obtain

$$K = \frac{\left(q_{Na_3AlF_6} + 2q_{Al}\right)\sqrt{2\pi R}}{S\sqrt{T}} \left[\frac{\int_0^\tau \left(I_{\frac{NaAlF_3^+}{NaAlF_4}} + I_{\frac{AlF^+}{NaAlF_4}} \right) \mathrm{d}t}{\sigma_{NaAlF_4} \cdot \sqrt{M_{NaAlF_4}}} + \frac{\int_0^\tau I_{\frac{AlF^+}{AlF}} \mathrm{d}t}{\sigma_{AlF}\sqrt{M_{AlF}}} + \frac{\int_0^\tau I_{Na^+/Na} \mathrm{d}t}{\sigma_{Na}\sqrt{M_{Na}}} \right] \tag{3.18}$$

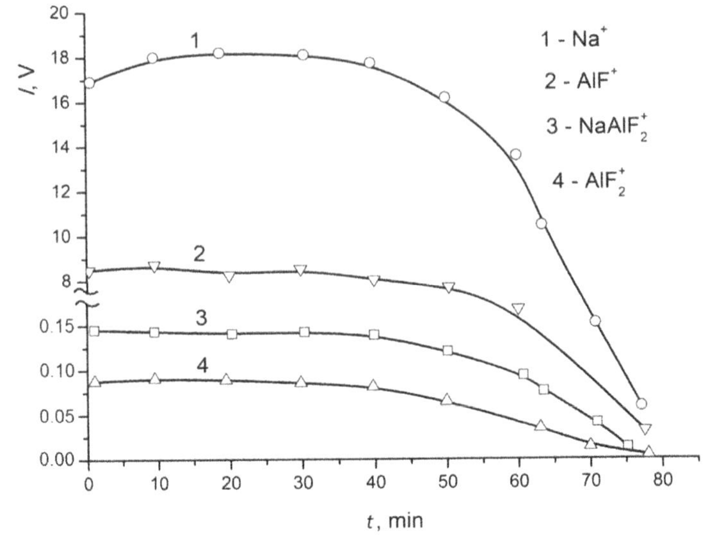

Fig. 3.6 Changes in the absolute intensities of ion currents of evaporation products of the Al/Na_3AlF_6 system with time. Samples: Na_3AlF_6: 15 mg, Al: 10 mg

As follows from the calculations, $K = (9.0 \pm 0.6) \times 10^{-25}$ atm cm^2/K V for the TiB_2 cell.

Table 3.5 shows the composition of saturated vapor over the Al/Na_3AlF_6 system.

However, the materials of the cells used were not inert. Thus, after the experiments, a black coating was found on the wall of the platinum chamber. Its diffraction pattern contains a set of reflections corresponding to Pt_2Al and PtAl intermetallics. Al_4C_3 was formed in the graphite chamber. Formally, these interactions were taken into account by means of a decrease in the activity of aluminum. We compared the

Table 3.5 Composition of the saturated vapor in Al/Na_3AlF_6 system and activities of its components

Cell material	Partial pressures (Pa)		
	Na	AlF	$NaAlF_4$
Platinum	0.71 ± 0.23	0.29 ± 0.02	1.08 ± 0.12
Graphite	0.94 ± 0.39	0.88 ± 0.18	1.34 ± 0.33
TiB_2	1.43 ± 0.24	1.09 ± 0.08	1.22 ± 0.15

Cell material	Total pressure (Pa)	Activities (mol fraction)		
		Na_3AlF_6	NaF	Al
Platinum	2.08 ± 0.37	1	0.057 ± 0.016	0.12 ± 0.05
Graphite	3.16 ± 0.90	1	0.052 ± 0.022	0.57 ± 0.42
TiB_2	3.74 ± 0.47	1	0.062 ± 0.011	0.96 ± 0.29

Note Percentage composition calculated from arithmetic mean of partial pressures

experimentally established values with those calculated from thermodynamic data. Equations (3.1), (3.2) and (3.19) with equilibrium constants $K_{p(i)}$ were used for the analysis:

$$Na_3AlF_6 \leftrightarrows 2NaF + NaAlF_4 \tag{3.19}$$

Then the corresponding activities will be equal to

$$a_{NaF} = \left[\frac{p_{Na}{}^2 \cdot p_{NaAlF_4} \cdot K_p(2)}{p_{AlF} K_p{}^2(1) \cdot K_p(19)}\right]^{1/3} \tag{3.20}$$

$$a_{Al} = \left[\frac{p_{AlF}{}^2 \cdot p_{Na} \cdot p_{Na_3AlF_6}{}^{1/3} \cdot a_{NaF}{}^{2/3}}{K_p{}^{1/3}(19) \cdot K_p(2) \cdot K_p(1)}\right]^{3/5}. \tag{3.21}$$

The obtained values are presented in Table 3.5. The values of a_{NaF} in cryolite crystals are higher than those reported in [6] ($a_{NaF} = 0.029$; $T = 1100$ K). When using platinum and graphite cells, as expected, due to the formation of chemical compounds Pt_2Al, PtAl and Al_4C_3, the activity of aluminum decreased $a_{Al} < 1$, which led to a change in the composition of vapor, while in a cell made of TiB_2, which practically does not interact with aluminum [1–3], the activity of aluminum did not change $a_{Al} \approx 1$.

Comparison of the data in Table 3.5 with the composition of vapor for $T = 1300$ K shows that in the latter case, the content of Na ($P_{Na} = 8.5$ kPa) prevails in the gas phase over AlF and $NaAlF_4$ (p_{AlF} and p_{NaAlF_4} by 1 kPa) [2]. Most likely, this is caused by an increase in the activity of NaF in the melt by almost an order of magnitude due to the dissociation of cryolite [1–3].

3.5 System Al–Si/Na_3AlF_6

The measurements were carried out at a temperature of 1043 K, when the metal phase was liquid in the concentration range 0–30 mass% Si [19]. In addition to the Na^+, AlF^+, $AlF_2{}^+$, and $NaAlF_3{}^+$ ions typical for the Al/Na_3AlF_6 system, the mass spectra revealed SiF^+, $SiF_2{}^+$, $SiF_3{}^+$ and $SiF_4{}^+$. Their IEC is shown in Fig. 3.7.

Such a simple shape of the curves is characteristic of ions having one precursor each [6]. Since it is unlikely that all of these ions are formed from the corresponding fluorides, the mass spectrum of the most stable of them, which is SiF_4 tetrafluoride, was taken [20]. The spectrum of artificial SiF_4 obtained according to the procedure [21] contained the same ions with a similar IEC shape and the same appearance potentials. Hence, it follows that the SiF^+, $SiF_2{}^+$, $SiF_3{}^+$ ions are fragments from the electron impact of SiF_4 molecules, which is the main silicon-containing particle in the gas phase above the Al–Si/Na_3AlF_6 system.

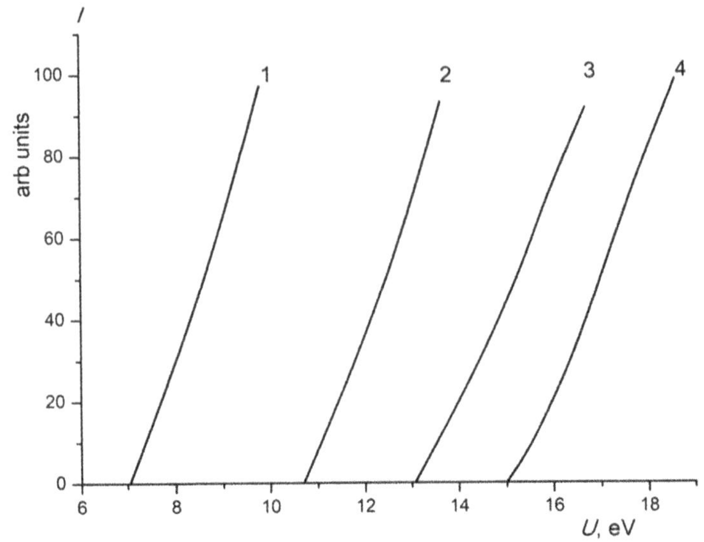

Fig. 3.7 Ionization efficiency curves for currents of silicon-containing ions. Designations: (1): SiF^+; (2): SiF_2^+; (3): SiF_3^+; (4): SiF_4^+

The interpretation of the remaining ionic currents was carried out, as in the case of the Al/Na_3AlF_6 system. The appearance of volatile SiF_4 makes quantitative measurements incorrect; therefore, a comparative estimation was made using an internal standard, which was taken as the total current of tetrafluoroaluminate $NaAlF_4$. Its molecules are formed only due to the evaporation of cryolite [6], and therefore, their partial pressure can be assumed to be constant. Figure 3.8 shows the dependences of the ratio of the ionic currents of Na, AlF and SiF_4 to the current of $NaAlF_4$ on the composition of the alloy. Equations (3.1) and (3.2) with equilibrium constants $K_{p(i)}$ were used for interpretation. It is obvious that the vapor pressures, and hence, the ion currents will change as follows: $I_{AlF} = K_1 a_{Al}^{2/3}$ and $I_{Na} = K_2 a_{Al}^{1/3}$. Theoretical dependences, constructed from the aluminum activity values borrowed from [22], are plotted in Fig. 3.8 by solid lines.

Up to a concentration of 20 mass% Si in the alloy, the experimental data are close to the calculated ones. Then, positive deviations take place along with an increase in the total SiF_4 current. Apparently, due to the increase in the activity of silicon in the alloy, the following reactions develop:

$$2Na_3AlF_6 + Si = 6NaF + AlF + SiF_4, \quad \Delta G_{1073K} = 503.9\ \text{kJ/mol} \tag{3.22}$$

$$4NaF + Si = 4Na + SiF_4, \quad \Delta G_{1073K} = 9.7\ \text{kJ/mol} \tag{3.23}$$

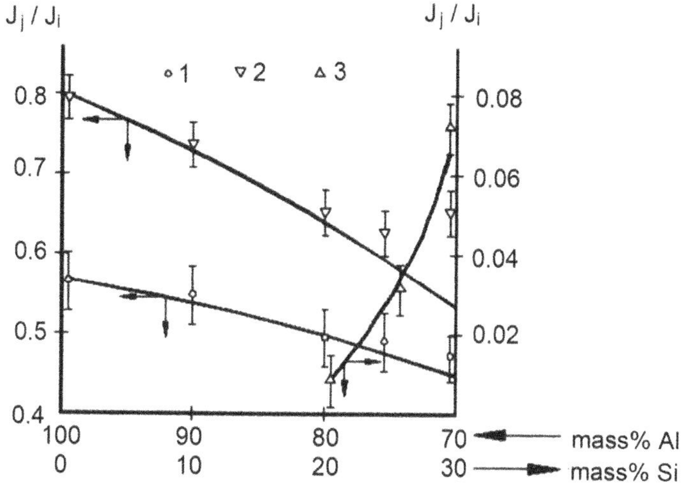

Fig. 3.8 Ratios of the ionic currents of Na, AlF and SiF_4 to the current of $NaAlF_4$. Designations: (1): I_{Na}/I_{NaAlF_4}; (2): I/I_{NaAlF_4}; (3): I_{SiF_4}/I_{NaAlF_4}

3.6 Solubility of Al–Si Alloy in Cryolite Melt

The solubility of the alloy in cryolite (c.r. = 3) was established by the method of isothermal saturation with subsequent treating the solidified salt with 10% hydrochloric acid and determining the amount of released hydrogen [7]. The scheme of the experimental setup is shown in Fig. 3.9.

The experiments were setting up as follows. The metal and salt reagents, in the amount of 10 and 80 g, respectively, were loaded into a boron nitride crucible 1, closed with a tightly ground lid 2 and placed in a reactor with a water-cooled stainless steel tower 3. The reactor was sealed, installed in a furnace with SiC heaters 4.6 kW power with a temperature control system, connected to water cooling and a vacuum-argon line. The retort was evacuated, filled with argon, and the gas was passed through at flow rate of 2.0–2.5 l/h. The vacuum was established by a VN-461M pump 5. Argon was cleaned from traces of moisture and oxygen by passing through a retort with a titanium sponge 6, and its consumption was controlled by a rotameter 7. Then the furnace was heated, the retort was thermostatically controlled and kept for 6 h. After that, without violating the tightness of the retort, the lid was removed from the crucible using a system of levers 8 and a sample of the melt was taken using a sampler 9. The temperature in the reactor was measured with a Pt/Pt–Rh thermocouple, the cold ends of which were in a Dewar vessel with melting ice, and recorded on a combined digital device 10 DD-2. After cooling in the tower, the solidified electrolyte sample was quickly transferred to a dry box filled with purified argon, ground in an agate mortar, and packed into an ampoule with a known weight. After weighing, the amount of salt loaded into the ampoule was determined by the difference, and the latter was attached to a volume meter with 10% hydrochloric

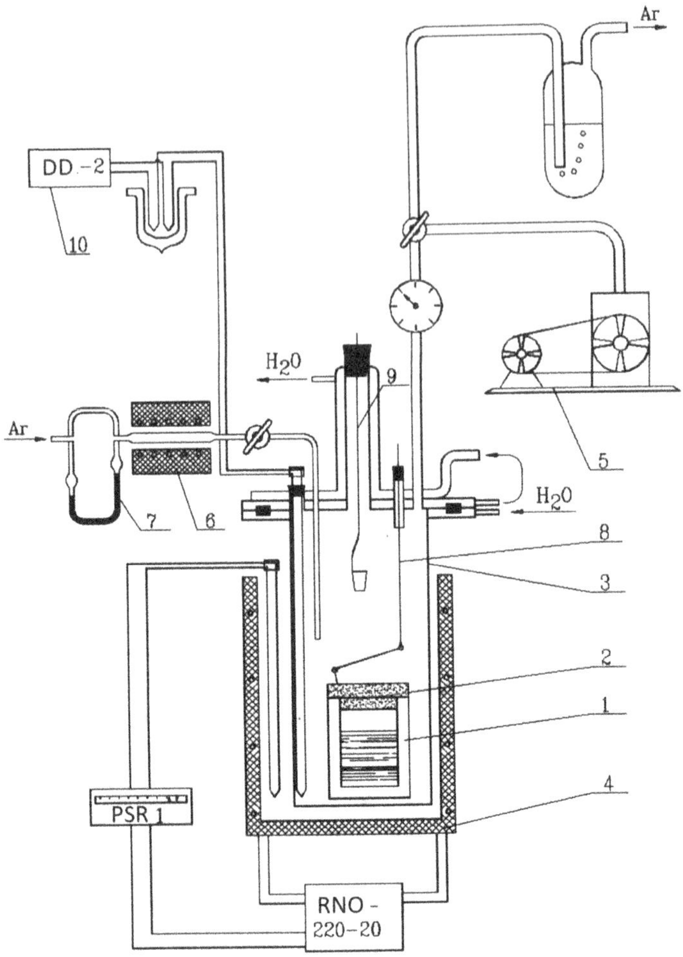

Fig. 3.9 Scheme of the installation for determining the solubility of Al–Si alloys. Designations: (1) reaction crucible; (2) cover; (3) reactor; (4) furnace; (5) vacuum pump; (6) argon purification reactor; (7) rotameter; (8) system of levers; (9) sampler; (10) combined digital device

acid saturated with hydrogen. At the end of the interaction, the amount of released hydrogen was measured and brought to normal conditions.

Under the dissolved aluminum, we understand the sum of the Na and $NaAlF_2$ in cryolite. During crystallization, the latter disproportionates, and at acid treatment of the sample, hydrogen is evolved by both sodium and aluminum. It is impossible to determine how much gas was released from each of the reagents separately by volumetric method. Therefore, all the collected hydrogen was formally recalculated to an equivalent amount of aluminum. The data obtained are shown in Fig. 3.10 and indicate a decrease in solubility with an increase in the concentration of the alloy.

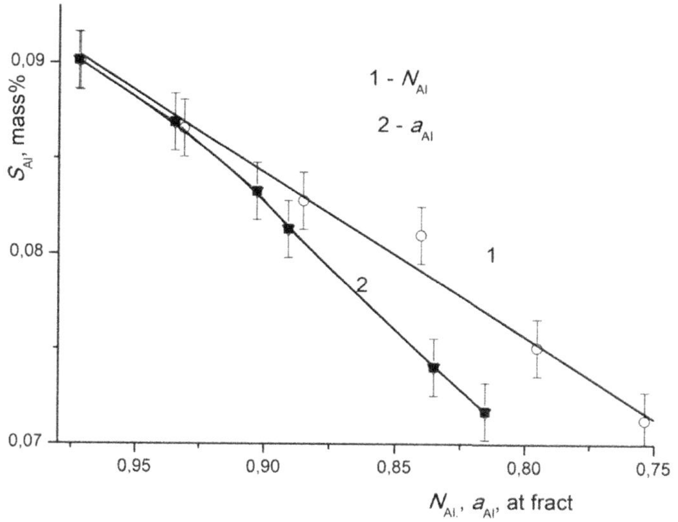

Fig. 3.10 Solubility of aluminum in Na_3AlF_6 melt depending on the concentration of Al–Si alloy. $T = 1300$ K

Let us determine the parts of Na and $NaAlF_2$ separately. For that, we write down the equilibrium constants of the reactions (3.3) and (3.4) via the activities:

$$K_p(3.3) = a_{NaAlF_2}{}^3 / a_{Al}{}^2 \cdot a_{Na_3AlF_6}, \tag{3.24}$$

$$K_p(3.4) = a_{Na}{}^3 \cdot a_{Na_3AlF_6} / a_{Al} \cdot a_{NaF}{}^6. \tag{3.25}$$

Taking into account that the range of concentration changes is not large, the activity coefficients γ_{Na} and γ_{NaAlF_2} can be considered to be constants (i.e., obeying the Henry law [23]), and the activities proportional to concentrations. Then the latters for Na and $NaAlF_2$ will be

$$C_{Na} = \left(K_p{}^{1/3}(3.4) a_{NaF}{}^2 / a_{Na_3AlF_6}{}^{1/3} \gamma_{Na}\right) a_{Al}{}^{1/3} \tag{3.26}$$

$$C_{NaAlF_2} = \left(K_p{}^{1/3}(3.3) a_{Na_3AlF_6}{}^{1/3} / \gamma_{NaAlF_2}\right) a_{Al}{}^{2/3} \tag{3.27}$$

And their sum will be equal to total solubility of aluminum:

$$C_{total} = C_{Na} + C_{NaAlF_2} = M a_{Al}{}^{1/3} + N a_{Al}{}^{2/3}, \tag{3.28}$$

where

$$M = K_p{}^{1/3}(3.4) a_{NaF}{}^2 / a_{Na_3AlF_6}{}^{1/3} \gamma_{Na} \tag{3.29}$$

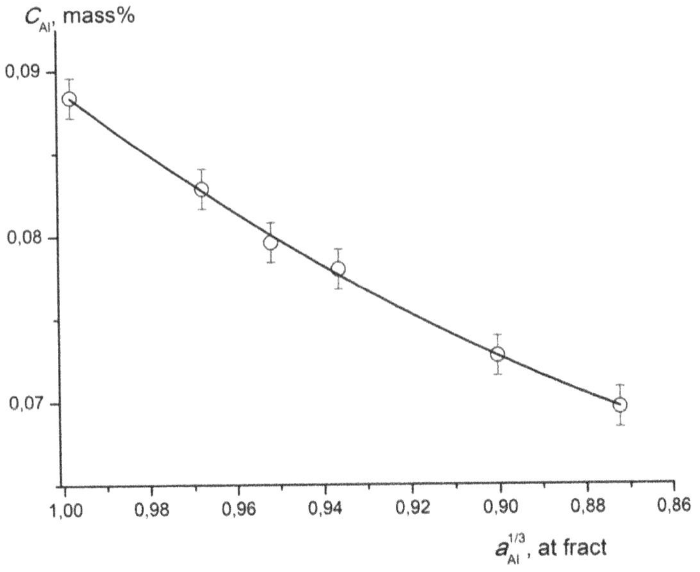

Fig. 3.11 Solubility of aluminum in Na_3AlF_6 melt represented as the function $S = f(a_{Al}{}^{1/3})$

and

$$N = K_p{}^{1/3}(3.3)a_{Na_3AlF_6}{}^{1/3}/\gamma_{NaAlF_2} \quad (3.30)$$

Determination of M and N is made by plotting of the graph $C_{total} = f(a_{Al}{}^{1/3})$ (Fig. 3.11) and its approximation by polynomial function $C_{total} = Mx + Nx^2$ ($x = a_{Al}{}^{1/3}$) [24].

Using the values of aluminum activity from [22], we obtain the function

$$C_{total} = 0.029 \cdot a_{Al}{}^{1/3} + 0.060 \cdot a_{Al}{}^{2/3}. \quad (3.31)$$

We can estimate the reliability of this function by comparing with the known literature data on the separate content of Na and $NaAlF_2$ during the dissolution of aluminum ($a_{Al} = 1$). Namely (in mass%), according to [7], they are 0.020 and 0.062, and, according to [8], 0.036 and 0.060, which are very close to our data 0.029 and 0.060. Thus, we can consider the obtained data on the solubility of Al–Si alloys in cryolite as quite adequate.

3.7 Conclusions

The study of the solubility of Al–Si alloys in cryolite, as well as the pressure and composition of the vapor, makes it possible to explain the shape of the graphs in

Figs. 3.1 and 3.2. The initial decrease in the evaporation rate of the Al–Si/Na_3AlF_6 system and metal losses with increasing alloy concentration are due to a decrease in its solubility in comparison with pure aluminum. Obviously, the intensity of AlF and Na volatilization depends on their content in cryolite and, therefore, will decrease. The presence of minima on the graphs at a content of 20 mass% Si in the alloy and a further increase in both the evaporation rate of the system and metal losses are caused by the separation of SiF_4 into the gas phase. However, such alloy concentrations are unattainable for the electrolytic method, and this situation is possible only when processing supereutectic alloys with cryolite-based fluxes.

Let us consider practical conclusions. The efficiency of the technology primarily depends on the change in the performance of the aluminum electrolytic cell when it is switched to the alloy production mode. In a quantitative assessment, we will proceed from opposite factors: the difference in electrochemical equivalents $E_{Si} = 0.2617$ g/Ah and $E_{Al} = 0.335$ g/Ah and a decrease in metal losses in the silicon concentration range from 0 to 10 mass% (it is real interval).

The reduction of a nobler element on a liquid cathode from a more active metal proceeds with a current efficiency close to the theoretical [9], which can also be expected for the electrodeposition of silicon on aluminum. The current efficiency of aluminum is determined by the loss rate of the dissolved metal, and, regardless of the loss mechanism (volatilization of AlF and Na, or their oxidation both electrochemically at the anode and by carbon oxides), this is a diffusion-controlled process [8].

As a result, the loss rate should be proportional to the concentration of dissolved aluminum [25], and then, obviously, the current efficiency should be inversely proportional to this value. So the performance of the electrolyzer (in terms of productivity change Δp) should be expressed as follows:

$$\Delta p = \frac{m_{Al-Si}}{m_{Al}} = \frac{\eta_{Si} E_{Si} Q M + \frac{\eta_{Al}}{K} \cdot E_{Al} Q (1 - M)}{\eta_{Al} E_{Al} Q}, \quad (3.32)$$

where η_{Si} and η_{Al}, E_{Si} and E_{Al} are current efficiencies and electrochemical equivalents, respectively, of silicon and aluminum; Q—electric charge passed; M—mass part of silicon in the alloy; K is the ratio

$$K = \frac{S_{Al-Si}}{S_{Al}} = \frac{0.029 a_{Al}{}^{1/3} + 0.060 a_{Al}{}^{2/3},\ \%}{0.089\%}, \quad (3.33)$$

where $S_{(i)}$ are solubilities of alloys and aluminum.

The calculations have shown that when obtaining alloys, for example, with 3 mass% Si, the productivity of the electrolyzer should increase by 1.013 times, and with 5 mass%, i.e., by 1.034 times. Therefore, the decrease in productivity observed during industrial tests of the method [26–29] was caused solely by an increase in the process temperature (up to 980 °C [26] and 1000 °C [29]), since its increase by 1° reduces the current efficiency by 0.3–0.7% [1–3].

References

1. Grjotheim K, Krohn C, Malinovsky M, Matiasovsky K, Thonstad J (1977) Aluminium electrolysis. Aluminium-Verlag, Dusseldorf, 350 p
2. Grjotheim K, Krohn C, Malinovsky M, Matiasovsky K, Thonstad J (1982) Aluminium electrolysis, 2nd edn. Aluminium-Verlag, Dusseldorf, 442 p
3. Thonstad J, Fellner P, Haarberg GM, Hives J, Kvande H, Sterten A (2001) Aluminium electrolysis, 3rd edn. Aluminium-Verlag, Dusseldorf, 359 p
4. Stokes J, Frank W (1963) Extractive metallurgy of aluminium, vol 2, pp 8–14
5. Belyaev A, Firsanova L (1959) Odnovalentnyi aliuminiy v menallurgicheskikh protsessakh (Univalent aluminum in metallurgical processes). Metallurgizdat, Moscow, 143 p (in Russian)
6. Sidorov L, Korobov M, Zhuravleva L (1985) Mass-spektralnye termodinamicheskiye issledovaniya (Mass-spectral thermodynamic studies). MGU, Moscow, 208 p (in Russian)
7. Odegard R, Sterten A, Thonstad J (1987) TMS light metals 1987, pp 389–396
8. Vetiukov M, Tsyplakov A, Shkolnikov S (1987) Elektrometallurgiya aliuminiya i magniya (Electrometallurgy of aluminum and magnesium). Metallurgiya, Moscow, 320 p (in Russian)
9. Morachevskiy A, Avaliani A, Mindin V (1978) Zhidkiye katody (Liquid cathodes). Mzinereba, Tbilisi, 184 p (in Russian)
10. Morozov I, Rykov A, Korenev Yu, Pruttskov D (1990) Rasplavy (Melts) 4:111–114 (in Russian)
11. Morozov I, Pruttskov D, Rykov A, Korenev Yu (1991) Rasplavy (Melts) 5:69–71 (in Russian)
12. Pruttskov D, Titaev P, Andriiko A, Mozhaev V, Polyakov P (1991) Rasplavy (Melts) 5(5):69–73
13. Pruttskov D, Titaev P, Olesov Yu (1991) Rasplavy (Melts) 5:108–110 (in Russian)
14. Suvorov A (1970) Termodinamicheskaya khimiya paroobraznogo sostoyaniya (Thermodynamic chemistry of vapours). Khimiya, Leningrad, 208 p (in Russian)
15. Gurvich L, Karachentsev G, Kondratiev V, Lebedev Yu, Medvedev V, Potapov V, Khodeev Yu (1974) Energii razryva khimicheskikh svyazei. Potentsialy ionizaysi i isrodstvo k elektrony (The energy of breaking chemical bonds. Ionization potentials and electron affinity). Nauka, Moscow, 351 p (in Russian)
16. Nikitin M, Sidorov L (1980) Int J Mass Spectrom Ion Phys 35:101–106
17. Stull D, Prophet H (1971) JANAF thermochemical tables, 2nd edn. U.S. Department of Commerce, National Bureau of Standards, 1141 p
18. Sterten A, Hamberg K, Maeland I (1982) Acta Chem Scand A 36:329–344
19. Stroganov G, Rotenberg B, Gershman G (1977) Splavy aliuminiya s kremniyem (Alloys of aluminum with silicon). Metallurgiya, Moscow, 271 p (in Russian)
20. Nekrasov B (1973) Osnovy obshchei khimii (Basics of general chemistry). Khimiya v.I., Moscow, 656 p (in Russian)
21. Rapoport F, Ilinskaya A (1963) Laboratornye metody polucheniya chistykh gazov (Laboratory methods for preparation of pure gases). Goskhimiszdat, Moscow, 420 p (in Russian)
22. Schaefer S (1974) Rept Invest Bur Mines US Dep Inter 7895:1–15
23. Smirnov M (1973) Elektrodnye potentsialy v pasplavlennykh khloridakh (Electrode potentials in molten chlorides). Nauka, Moscow, 248 p (in Russian)
24. Andriiko A, Andriiko Yu, Nauer G (2013) Many-electron electrochemical processes. Springer-Verlag, Berlin, 167 p
25. Frank-Kamenetskii D (2015) Diffusion and heat exchange in chemical kinetics. Princeton University Press, 384 p
26. Hannach R, Osborne J, Templeton G, Frazer E, Welch B (1977) Molten salts electrolysis in metal production. In: Proceedings of international symposium, Grenoble, pp 7–13
27. Tabereax A, McMinn C (1978) TMS light metals 1978, pp 209–222
28. Keller R, Welch B, Tabereaux A (1990) TMS light metals 1990, pp 333–340
29. Moxnes B, Gikling H, Kvande H, Rolseth S, Straumsheim K (2003) TMS light metals 2003, pp 329–334

Chapter 4
Industrial Tests

4.1 Introduction

There were several attempts to obtain Al–Si alloys by loading SiO_2 into industrial aluminum electrolyzers [1–5]. All authors encountered the same technological problems, namely deposit formation, and an increase in voltage and process temperature, and a decrease in current efficiency were observed. The cause of the negative phenomena had not been established then.

Based on our studies, we concluded that these problems are the result of the formation of a poorly soluble silicon compound in an intermediate oxidation state $[Si(II)]_{solid}$ at the cryolite-oxide melt/liquid cathode interface [6–8] (see Chap. 2). As a result, additional ohmic resistance appears which leads to an increase in voltage and temperature. Due to the intensive movement of liquid media in the electrolytic cell, the passivation layer is broken and part of it accumulates in the form of sludge under aluminum surface. This chapter presents the modes of maintenance of electrolyzers, which exclude the mentioned violations [9, 10].

4.2 Dissolution Rate of SiO_2-Containing Minerals in the Cryolite Melts

Precipitation can also be caused by differences in the dissolution rates of the raw materials. For alumina, this parameter has been studied repeatedly [11–13], but for quartz, it was done only once, by the method of a rotating disk [14]. The behavior of other SiO_2-containing substances has not been studied at all. The objects of the study were minerals that are mined in Ukraine with composition, mass%: quartz sand ($Fe_2O_3 < 0.04$; $Al_2O_3 < 0.6$); disthene-sillimanite concentrate (DSC) $Al_2O_3 \cdot SiO_2$ (Al_2O_3: 55–57; Fe_2O_3, ZrO_2, TiO_2: 0.7–0.8 each; CaO, MgO: ~ 0.3 each; SiO_2: rest) and kaolin $Al_2O_3 \cdot 2SiO_2 \cdot 2H_2O$ (SiO_2: 47–50; TiO_2: ~ 0.9; Fe_2O_3: ~ 1.2;

D. Pruttskov et al., *Electrolytic Production of Al–Si Alloys*,
Monographs in Electrochemistry, https://doi.org/10.1007/978-3-031-29249-1_4

CaO, MgO: ~ 0.2 each; Na_2O, K_2O: ~ 0.3 each; Al_2O_3: rest, LOI (loss on ignition): 13.5–14.0), and also alumina grade G-0 (GOST 6912).

The sand and DSC were dried at the processing plant. Kaolin contains the chemically bound water, which will react with cryolite (hydrolysis) and thus must be calcinated. We used dust captured from the exhaust gases of a rotary kiln in which kaolin clinker was produced. Its diffraction pattern showed reflections of mullite and α-SiO_2 which were formed by the reaction

$$3[Al_2O_3 \cdot 2SiO_2 \cdot 2H_2O] \rightarrow 3Al_2O_3 \cdot 2SiO_2 + 4SiO_2 + 6H_2O\uparrow \tag{4.1}$$

The fractional composition of the ingredients was determined on a ROTAP machine (sand, DSC) and on a Kvantimet-720 device (alumina, kaolin). Conditionally assuming the spherical shape of the particles and using the density values (g/cm^3: sand—2.6; kaolin—2.7; DSC—3.2; alumina—3.9 [15]), we found their specific surface area in cm^2/g.

The measurement technique consisted in visual observation of the disappearance of a substance in a cryolite melt during the bubbling of argon through it [16]. The experiments were carried out in a platinum crucible with dimensions $d = 50$ mm and $h = 100$ mm, in which there was 100 g of the melt. Argon was supplied through a platinum tube placed in the center of the bottom of the crucible, at a rate of 2 l/h.

When an ingredient enters the melt, its "cake" is initially formed, which is obviously associated with the freezing of salt on cold particles of powders. Due to the bubbling of gas, the "cake" does not sink to the bottom of the crucible, but, as it were, floats in the bulk of the melt. Over time, it breaks up into small conglomerates, which then dissolve in cryolite. The measurement results are summarized in Table 4.1.

The dissolution rate of alumina in cryolite is somewhat higher than that for sand and aluminosilicates. However, the dosage of the latter into the electrolytic cell is

Table 4.1 Results of measurements of the rate of dissolution of raw materials in the melt Na_3AlF_6, $T = 1303$ K

Material	Average particle diameter (mm)	Specific surface area (cm^2/g)	Mass of sample (% from the mass of melt)	Time of dissolution (s)	Rate of the dissolution (g/cm^2 s × 10^{-4})
1	2	3	4	5	6
Alumina	0.029 ± 0.003	930 ± 50	0.5	12 ± 2	1.59 ± 0.35
			6.0	148 ± 8	1.57 ± 0.23
Sand	0.216 ± 0.023	110 ± 12	0.5	30 ± 4	1.46 ± 0.33
			1.0	63 ± 6	1.44 ± 0.29
DSC	0.120 ± 0.010	160 ± 14	0.5	23 ± 3	1.35 ± 0.27
			1.0	48 ± 5	1.32 ± 0.24
			0.5	3 ± 1	1.12 ± 0.60
Kaolin	0.015 ± 0.004	1480 ± 50	1.0	7 ± 1	0.96 ± 0.33

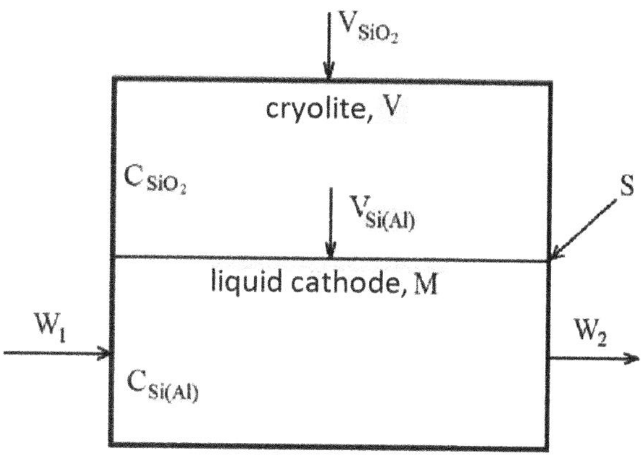

Fig. 4.1 Structure of the system in the form of two combined ideal mixing reactors. Designations: (V) volume of cryolite; (S) and (M) are the area and mass of the liquid cathode; (W_1) productivity; (W_2) metal extraction; (V_{SiO_2}) SiO_2 input rate; ($V_{Si(Al)}$) the rate of transition of Si into the liquid cathode; (C_{SiO_2}) content of SiO_2 in cryolite; ($C_{Si(Al)}$) alloy concentration. See explanations in the text

much less by weight than alumina, and, therefore, the dissolution time of SiO_2-containing minerals will be less than that for alumina.

4.3 Mathematical Model of Al–Si Alloy Production in Aluminum Electrolyzer

4.3.1 *Ideal Mixed Reactor Model*

In the aluminum electrolyzer, strong circulation of both the cryolite-oxide melt and the liquid cathode is observed. Therefore, the production of Al–Si alloys in it can be represented as a process taking place in two combined ideal mixing reactors (Fig. 4.1). In chemical technology, this is understood as a reservoir in which, due to intensive mixing (e.g., a vessel with a propeller stirrer), the concentrations of substances are instantly averaged [17].

4.3.2 *Macrokinetics of Silicon Reduction*

To develop an optimal technology for production of Al–Si alloys, it is necessary to know the Si mass transfer coefficient in the cryolite-oxide melt $K_m = D/\delta$ (where D is the diffusion coefficient; δ is the diffusion layer thickness [18]), which value depends on the hydrodynamic situation in the electrolytic cell. Therefore, the macrokinetics of Si reduction was studied in industrial Soderberg units with a current of 67 kA, in which the area of the Al cathode is $S = 9.0 \pm 0.5$ m^2, and the volume of cryolite is $V = 1.2 \pm 0.1$ m^3. A sample of sand weighing 20 kg was loaded into windows in the electrolyte crust, punched along the long side of the electrolyzer. Its dissolution and averaging of the composition took place within 15–20 min. Then, every hour,

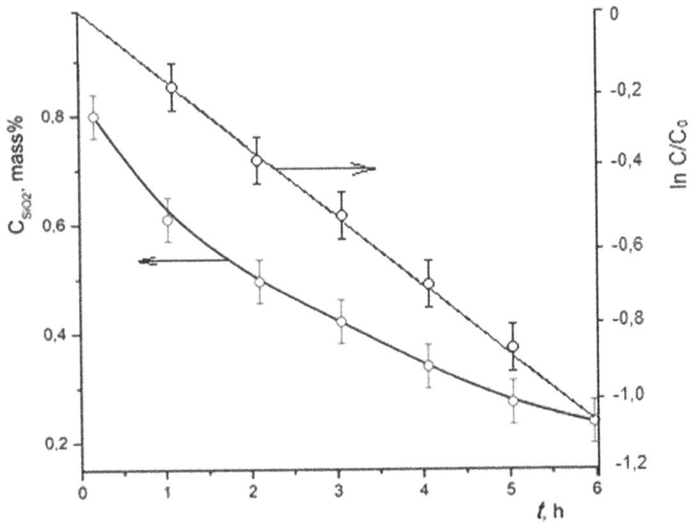

Fig. 4.2 Dynamics of change in the content of SiO_2 in the cryolite of the electrolyzer for a current of 67 kA

samples of cryolite were taken at eight points along the perimeter of the aggregate and the content of SiO_2 in them was determined (GOST 10561). Based on the data obtained, a graph was built (Fig. 4.2), which is approximated by an exponential dependence:

$$C = C_0 \exp[-t/\tau], \tag{4.2}$$

where t is time and $\tau = 19{,}320$ s.

This slope is linear in the coordinates of Eq. (4.2), which additionally confirms the diffusion control of the silicon reduction process [19].

It is known that the concentration change of a substance in a heterogeneous process limited by linear diffusion is described by equation [19]

$$C = C_0 \exp[(-K_m \cdot S/V)t]. \tag{4.3}$$

Thus, knowing S and V, we have calculated the mass transfer coefficient of Si, $K_{m(Si)} = (0.69 \pm 0.10) \times 10^{-5}$ m/s.

The Al mass transfer coefficient for acid fluoride melt (~ chiolite) obtained under conditions of free convection in a laboratory cell was $K_{m(Al)} = 7.0 \times 10^{-5}$ m/s [20], which is significantly higher than that for Si in an industrial electrolytic cell, where forced convection takes place. This indicates that complex aluminum ions are much more mobile than charged silicon particles existing in the form of bulky nepheline-like cycles [21, 22] (see Chap. 1). Obviously, both the low value of the mass transfer coefficient of Si and the limited solubility of its intermediate compound

$[Si(II)]_{solid}$ [6–8] in cryolite (see Chap. 2) will be factors preventing the production of concentrated Al–Si alloys in the electrolytic cell.

4.4 Attainable Concentration of Silicon in the Electrolytic Alloy

The condition for the formation of a poorly soluble intermediate $[Si(II)]_{solid}$ is the passivation criterion [6–8]

$$PC_{Si(II)} = \left[C_{SiO_2}\right]\left\{a_{Si(Al)}\right\} \geq (1.27 \pm 0.25) \times 10^{-4}, \tag{4.4}$$

where C_{SiO_2}—concentration of SiO_2 in cryolite; $a_{Si(Al)}$—activity of Si in Al–Si alloy (see Chap. 2).

Let us calculate the theoretically achievable concentration of Si in the alloy when $[Si(II)]_{solid}$ precipitate does not yet appear. Obviously, the problem is reduced to constructing the dependence

$$\left[C_{SiO_2}\right]\left\{a_{Si(Al)}\right\} = f\left\{C_{Si(Al)}\right\}, \tag{4.5}$$

where $C_{Si(Al)}$—concentration of the alloy.

Two methods of loading SiO_2 into the electrolytic cell were considered, namely (a) periodic mode, in which SiO_2 is fed at regular intervals to the electrolyte crust with its subsequent destruction; (b) continuous mode, in which SiO_2 is introduced directly into the cryolite using automatic feeders.

In periodic mode, the concentration of Si in the alloy can be calculated by the equation

$$C_{Si(Al)} = \frac{E_{Si}Q_{Si}\eta_{Si}}{E_{Al}(Q_{total} - Q_{Si})\eta_{Al} + E_{Si}Q_{Si}\eta_{Si}} \times 100\% \tag{4.6}$$

where E_{Si} and E_{Al}, η_{Si} and η_{Al}—electrochemical equivalents and current efficiencies of Si and Al; Q_{total} and Q_{Si}—charge passed through the electrolyzer and consumed for the reduction of Si.

Q_{Si} is determined by integrating the equation which describes the current value I_d for the reduction of silicon compounds

$$I_d = 4FK_mC_{SiO_2}S \cdot \exp\left(-K_m \cdot \frac{S}{V} \cdot t\right) \tag{4.7}$$

where F—Faraday number, K_m—mass transfer coefficient, V—volume of cryolite, S—the surface area liquid cathode, C_{SiO_2}—concentration of SiO_2 in cryolite, t—time.

The integration is carried out from zero to the time of the next loading of raw materials. Since the electrolyte crust is destroyed every nth hours, then

$$Q_{\rm Si} = 4FK_{\rm m}C_{\rm SiO_2}S\int_0^n \exp\left(-K_{\rm m}\cdot\frac{S}{V}\cdot t\right){\rm d}t$$
$$= 4FC_{\rm SiO_2}V\left[1-\exp\left(-K_{\rm m}\cdot\frac{S}{V}\cdot n\right)\right] \quad (4.8)$$

Functional relation $a_{\rm Si(Al)} = f(C_{\rm Si(Al)})$ was built using the data [23]. In the range of $C_{\rm Si(Al)} = 0.01$–0.15 mass fractions, it is described by the equation

$$a_{\rm Si(Al)} = 1.6824C^2{}_{\rm Si(Al)} - 0.0224C_{\rm Si(Al)} + 0.0022 \quad (4.9)$$

We accept the following conditions: Si as a noble element is discharged on an Al cathode with a theoretical current efficiency ($\eta_{\rm Si} = 1$) [24]; $\eta_{\rm Al} = 0.87$ and $n =$ 4 h. The graph of function (4.5) was found by numerically solving the system of Eqs. (4.6), (4.8) and (4.9). It is shown in Fig. 4.3 (curve 1).

In a continuous mode of loading, which provide a constant content of SiO_2 in cryolite, the equation for calculating the concentration of the alloy has the form

$$C_{\rm Si(Al)} = \frac{i_{\rm d}E_{\rm Si}\eta_{\rm Si}}{(i_{\rm c}-i_{\rm d})E_{\rm Al}\eta_{\rm Al} + i_{\rm d}E_{\rm Si}\eta_{\rm Si}} \times 100\%, \quad (4.10)$$

where $i_{\rm c}$ and $i_{\rm d}$—total cathodic and limiting for silicon current densities.

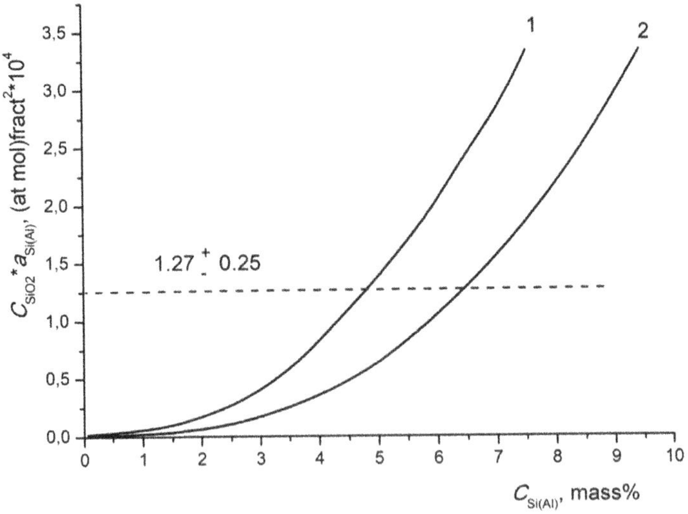

Fig. 4.3 Dependence of $[C_{\rm SiO_2}]\{a_{\rm Si(Al)}\}$ on $C_{\rm Si(Al)}$ in different modes of feeding the electrolyzer. Designations: (1) periodic; (2) continuous

Taking i_c ~ 0.7 A/cm^2, and i_d ~ 0.05 A/cm^2 (at C_{SiO_2} ~ 0.5 mass%) and making similar calculations, we obtain graph of the function (4.5), which is shown in Fig. 4.3 (curve 2).

As follows from the data, it is possible to obtain alloys with concentrations, mass% Si: in a periodic mode it is 5 ± 0.5; with continuous feeding—7 ± 0.5. When trying to increase the concentration of the alloy by increasing the load of SiO_2 into the electrolytic cell, an excess amount of Si in the intermediate oxidation state $[Si(II)]_{solid}$ will precipitate, which will block the surface of the liquid cathode, and the technology will be destabilized.

4.5 Transition Period of Operation

In periodic operation mode, the content of SiO_2 in cryolite will change as follows (Fig. 4.4). The maximum value C'_{max} can be described by ith order polynomial function

$$C'_{max} = \Delta C_{SiO_2}\left(a^i + a^{i+1} + \cdots + a + 1\right), \quad (4.11)$$

where i is the number of downloads of raw materials; ΔC_{SiO_2}—concentration change after loading; $a = \exp\left(-K_m \frac{S}{V} t\right)$.

Let us determine the number of downloads of raw materials, after which the change in concentration of SiO_2 (C'_{max}) in the electrolyte can be neglected. Mathematically, this condition is formulated as follows:

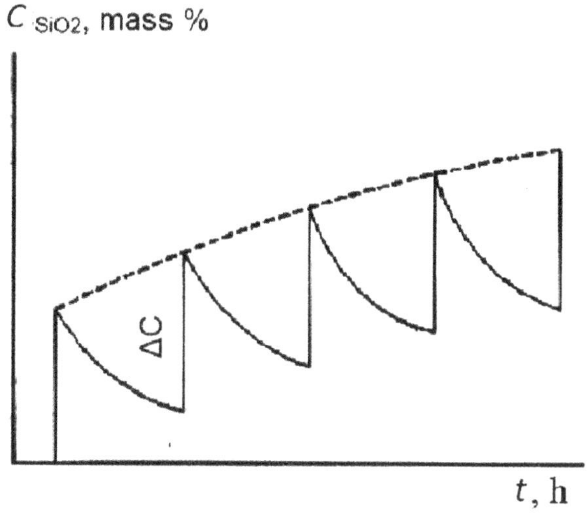

Fig. 4.4 Pattern of the change of SiO_2 content in cryolite (initial time period) at the periodic loading of raw material into the electrolyzer

$$\frac{a^i + a^{i+1} + \cdots + a + 1}{a^{i-1} + a^{i-2} + \cdots + a + 1} < \varepsilon \tag{4.12}$$

Since both the numerator and denominator of (4.12) are geometric progressions, the problem of finding i is reduced to determining the ratio of their sums

$$i > \frac{\ln \frac{\varepsilon - 1}{\varepsilon - a}}{\ln a}. \tag{4.13}$$

Given, for example, $\varepsilon = 1.05$, which means that each subsequent load does not lead to an increase in the concentration of SiO_2 in the electrolyte by more than 1.05 times and, if a time between loads is 4 h, the exponent value is $a = \exp(-5.175 \times 10^{-5} \times 4 \times 3600) = 0.475$. Then from (4.13), the number of downloads should be more than 4. It means that the induction period after which a rapid increase in the silicon concentration in the alloy should begin is $i \cdot n = 16$ h for an electrolytic cell of this type.

At the periodic maintenance of electrolyzers, the main parameters of the technology are the mass of a single portion of SiO_2 (Δm) and the time interval between the destruction of the electrolyte crust. Reducing the latter does not seem appropriate, since labor costs will increase significantly.

The value of Δm is calculated using Eqs. (4.4) and (4.11). According to Fig. 4.2, it is possible to obtain alloys containing up to 5 ± 0.5 mass% silicon in this mode. From (4.4), the concentration of SiO_2 in cryolite should be ~ 0.57 mass%. Then, from (4.11), it follows

$$0.57\,\text{mass\%} = \Delta C_{SiO_2}\left(a^i + a^{i+1} + \cdots + a + 1\right) = \Delta C_{SiO_2}\left(\frac{1 - a^{j+1}}{1 - a}\right) \tag{4.14}$$

Since in our example $0 < a < 1$ ($a = 0.475$), at $i \to \infty$ Eq. (4.14) will take the form

$$0.57\,\text{mass\%} = \Delta C_{SiO_2}/(1 - a) = \Delta C_{SiO_2}/0.525 \tag{4.15}$$

Or

$$\Delta C_{SiO_2} = 0.57\,\text{mass\%} \times 0.525 = 0.30\,\text{mass\%}$$

Then

$$\Delta m = \Delta C_{SiO_2} \cdot M \tag{4.16}$$

where M is the mass of cryolite-oxide melt.

In our case $V = 1.2 \pm 0.1$ m^3, and since $M = Vd$ (d ~ 2.07 t/m^3 [11–13]), we obtain the maximum single portion of SiO_2 (Δm) ~ 8 kg for the time interval 4 h between loads.

In a continuous loading mode, the time to reach a stationary concentration of C_{SiO_2} in cryolite is determined by the material balance equation (Fig. 4.1)

$$\frac{dC_{SiO_2}}{dt} = A - BC_{SiO_2}, \tag{4.17}$$

where $A = V_{SiO_2}/V$; $B = K_m \cdot S/V$; V_{SiO_2}—the rate of SiO_2 flow.

Separating the variables and integrating by changing the variable, we obtain the equation

$$C_{SiO_2} = A/B\left[1 - \exp(-Bt)\right] \tag{4.18}$$

The analysis of this equation shows that, for such type of electrolyzers, the time to reach 95% of the SiO_2 content in the electrolyte from the specified one (determined by the ratio A/B) is, as in the case of a periodic loading mode, 16 h.

The dynamics of silicon content increase in the alloy also obeys the material balance equation (Fig. 4.1). We accept that the quantities of produced and extracted metal are equal $W_1 = W_2 = W$, and, therefore, its amount in the electrolyzer remains constant $M =$ const, and then

$$M\frac{dC_{Si(Al)}{}^{(2)}}{dt} = W\left[C_{Si(Al)}{}^{(1)} - C_{Si(Al)}{}^{(2)}\right], \tag{4.19}$$

where $C_{Si(Al)}{}^{(1)}$ and $C_{Si(Al)}{}^{(2)}$ are the preset and current concentrations, respectively.

Separating the variables and integrating by changing the variable, we find the desired relation

$$C_{Si(Al)}{}^{2} = C_{Si(Al)}{}^{1}\left[1 - \exp\left(-\frac{W}{M}t\right)\right] \tag{4.20}$$

As follows from (4.20), the increase of current concentration of $C_{Si(Al)}{}^{(2)}$ is exponential and depends on the W/M ratio. For the electrolyzer of 67 kA current load, W ~ 470 kg/day and M ~ 5000 kg. Then the condition $C_{Si(Al)}{}^{(2)} = 0.95C_{Si(Al)}{}^{(1)}$ will be met in about a month.

Dynamics of $C_{Si(Al)}{}^{(2)}$ increase for other electrolyzers may not be the same because of the differences in W/M values.

4.6 Analysis of the Early Results of Industrial Tests of the Technology

Apparently, the first attempt to obtain an Al–Si alloy in an electrolyzer for a current of 23 kA was made in the 1930s [1]. It was reported that the alloys containing 10.6–13.0 mass% Si were obtained. The technological process was accompanied by an increase in temperature and the appearance of a dense precipitate. The loading of Al_2O_3 and SiO_2 on the electrolyte crust was carried out together, although the details of maintenance and, in particular, the dosage of SiO_2 and the time interval between the destruction of the crust were not reported. However, it is known from another source [25] that, according to the technology adopted at that time, alumina was supplied to the unit only after the occurrence of the anode effect. To reduce the number of the anode effects, the portion of the loaded raw materials was increased, which made it possible to extend the time between "flashes" to 16–20 h. Possibly, similar maintenance of the experimental electrolyzer was used, which naturally led to an overdose of SiO_2, and, therefore, to passivation phenomena and violations of technology.

The results of testing the method in an electrolyzer for a current of 85 kA are presented in [2]. It was reported that alloys containing 6.6 mass% Si were obtained by loading 18 kg of sand (0.5% of the cryolite weight) onto the electrolyte crust and destroying the crust every 4 h. A decrease in the current yield, an increase in temperature to 980 °C and the formation of a precipitate were noted. Knowing the unit load of sand, and also assuming that the ratio *S*/*V* and the conditions of mass transfer in this unit are the same as for the current of 67 kA, we find the concentration of SiO_2 in cryolite according to Eq. (4.11) $C'_{max} = 0.95$ mass%. Then, the product $[C_{SiO_2}]\{a_{Si(Al)}\} = (3.3 - 3.8) \times 10^{-4}$, which is larger than allowed (4.4), and the poorly soluble precipitate of $[Si(II)]_{solid}$ should form.

The data obtained in the electrolyzer for a current of 80 kA are given in [4]. Quartz sand was loaded onto the electrolyte crust every 4 h only in the daytime in four portions. At a dosage of sand in the amount of 1% of the mass of the melt, the actual content of SiO_2 in cryolite was 1.5–2.0 mass%, which is close to that calculated by Eq. (4.11). The concentration of alloy was only $C_{Si(Al)} = 5.0$–6.0 mass%. Then, $[C_{SiO_2}]\{a_{Si(Al)}\} = (5 - 7) \times 10^{-4}$, and the passivation should occur. Really, sludge formation, an increase in temperature and voltage was observed, and the productivity of the unit decreased.

The detailed report was published on the production of Al–Si alloys by loading crushed spent pot lining into an electrolyzer for a current of 115 kA in 2003 [5]. Its loading was carried out only once a day through a broken electrolyte crust. Already at the first stage, when only 80 kg (42 kg of SiO_2) of raw materials were loaded at a time, the temperature increased to 1000 °C, which required stopping the supply of material and shunting the bath. Subsequently, a one-time portion was increased to 240 kg (130 kg for SiO_2) and then to 360 kg (195 kg for SiO_2). The electrolyzer worked in such modes for 40 and 20 days, producing alloys with 8 and 11 mass% silicon, respectively, to the end of each period. Immediately after loading, the actual content of SiO_2 in cryolite was 0.75 and 1.0 mass%, which was three times less than expected. The authors attributed this observation to the slow "melting/dissolution" of the material. Under these conditions, the product $[C_{SiO_2}]\{a_{Si(Al)}\}$ was 2.83×10^{-4} and 6.82×10^{-4}, respectively, which is much higher than the admissible (4.4) and therefore a poorly soluble compound $[Si(II)]_{solid}$ was formed. Obviously, the unbalance in SiO_2 was associated with this phenomenon. The temperature was higher than usual by 24 °C, which required periodic termination of the loading of this raw material and shunting the unit.

Summarizing, we should admit that all attempts to implement the method of electrolytic production of Al–Si alloys on an industrial scale failed due to incorrect dosage of SiO_2.

4.7 Studies and Development of Industrial Technology on Söderberg Electrolyzers with 67 kA Current Load

4.7.1 *Loading the Sand onto the Crust of the Electrolyte*[1]

Sand was loaded onto the electrolyte crust regularly from the long sides of the unit 6 times a day every 4 h, followed by the destruction of the crust and the supply of alumina. The initial portion of sand was $\Delta m = 4$ kg, and after a month, it was increased to $\Delta m = 8$ kg. The dynamics of change in silicon concentration in the alloy is shown in Fig. 4.5. There is also presented the part of the $C_{Si(Al)}$–t curve corresponding to a sand dose of 48 kg/day, according to Eq. (4.20), in the coordinates $\ln(C_{Si(Al)}{}^{(1)} - C_{Si(Al)}{}^{(2)})$–$t$.

The linearity of the graph confirms the validity of the mathematical model based on the ideal mixing reactor. There was no deterioration in the operation of electrolyzers: the temperature and operating voltage remained at the same level, and the content of SiO_2 in cryolite was ~ 0.5 mass%.

Then the dose of sand was increased to 72 kg/day. A few days later, complications began: the temperature and voltage increased, a precipitate appeared. The concentration of SiO_2 in cryolite reached ~ 0.9 mass%, and the product $[C_{SiO_2}]\{a_{Si(Al)}\} = 2.6 \times 10^{-4}$, which resulted in the formation of $[Si(II)]_{solid}$ compound. After reducing the dosage to the initial value of 48 kg/day, the electrolytic cell worked stably for a year, producing alloys with ~ 5 mass% Si. The obtained values of the maximum portion of SiO_2 (Δm) and the achievable concentration of the $C_{Si(Al)}{}^{(1)}$ alloy completely coincided with the calculated ones.

4.7.2 *Continuous Loading of the Sand*

At the next stage, the sand was loaded directly into the cryolite melt using two automatic dispensers located on the end sides of the electrolyzer, where there were

---

[1] The industrial tests of the technology had started in the summer of 1982. The experimental electrolyzer at the end of its service life was chosen. "Let them take this one, no harm, anyway it is dying"—said the foreman of electrolysis department. So, we christened this unit as "The Dying Swan".

The first USSR patent was filed a year later (No. 1115490, 22.05.84).

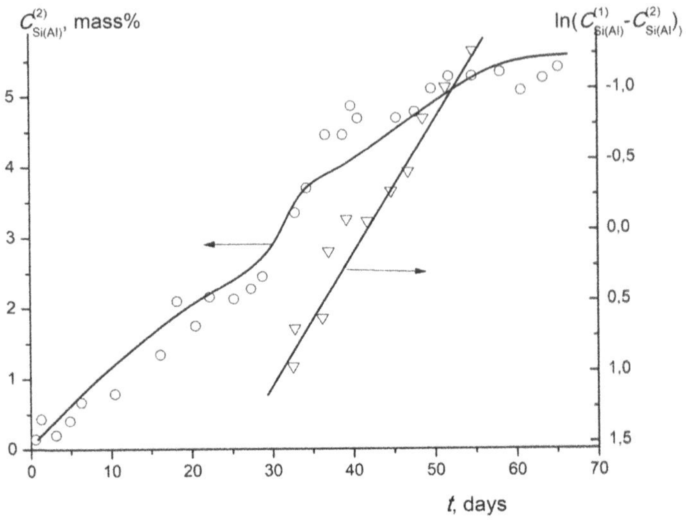

Fig. 4.5 Graph of the increase in the concentration of Si in the alloy and its presentation in coordinates corresponding to the ideal mixing reactor

no anode pins. Examples of graphs describing the time variation of the SiO_2 content in cryolite are shown in Fig. 4.6. The steady state is reached, as predicted by Eq. (4.18), in 16 h, which additionally indicates the applicability of the ideally mixed reactor model.

Examples of the dependences for change in time of the Si content in the alloy (after reaching a stationary concentration of SiO_2 in cryolite), taking into account the daily production of metal by the electrolyzer, are shown in Fig. 4.7.

From the tangent of the slope of the straight lines $C_{Si(Al)}-\tau$, we have calculated the rate of Si transport from salt phase to the metal $V_{Si(Al)}$. It turns out to be equal to the rate of feeding of Si into cryolite in the form of the oxide $V_{Si(Al)} = V_{SiO_2}$, if the values of V_{SiO_2} do not exceed 14×10^{-4} mol/m^2 s (curve 1, Fig. 4.8).

In the range of the oxide input rates $V_{SiO_2} = (14\text{–}16) \times 10^{-4}$ mol/m^2 s, an inflection was observed on curve 1. Further increase in V_{SiO_2} did not lead to a corresponding increase in $V_{Si(Al)}$. The graph describing the change in the stationary concentration of SiO_2 in cryolite also goes symbatically (curve 2, Fig. 4.8).

After the end of the linear part, the values of the SiO_2 content in cryolite were poorly reproduced, the formation of a precipitate was observed, as well as an increase in temperature and voltage. At $V_{SiO_2} = (14\text{–}15) \times 10^{-4}$ mol/m^2 s or 65–70 kg/day, the electrolyzer worked stably, producing alloys with 7.5–8.0 mass% Si.

Let us interpret the kinetics of Si reduction based on the existence of intervalence equilibrium with the constant K_{in}

$$[Si(IV)] + [Si(Al)] \leftrightarrows 2[Si(II)]. \quad (4.21)$$

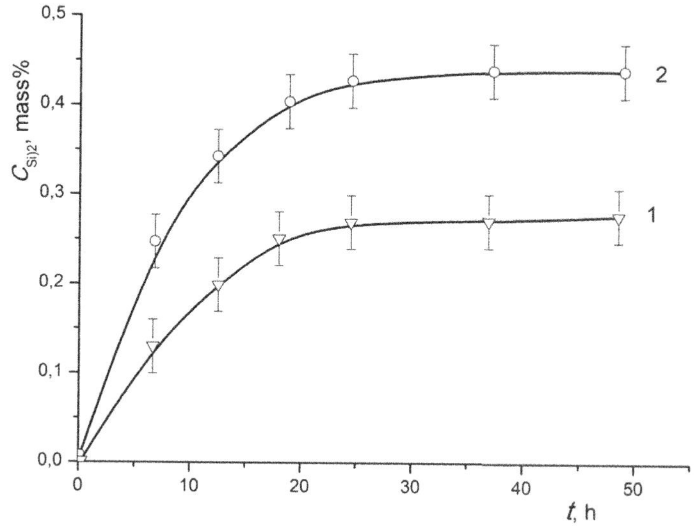

Fig. 4.6 Examples of time dependences for changes of SiO_2 concentration in cryolite. Designations: (1) $V_{SiO_2} = 6 \times 10^{-4}$ mol/m^2 s; (2) $V_{SiO_2} = 1 \times 10^{-3}$ mol/m^2 s

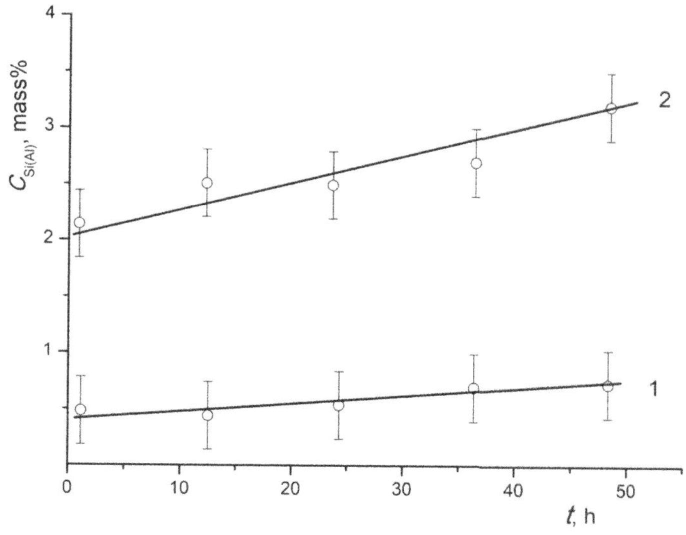

Fig. 4.7 Examples of time dependences of Si content in the alloy. Designations: (1) $V_{SiO_2} = 4 \times 10^{-4}$ mol/m^2 s; (2) $V_{SiO_2} = 8 \times 10^{-4}$ mol/m^2 s

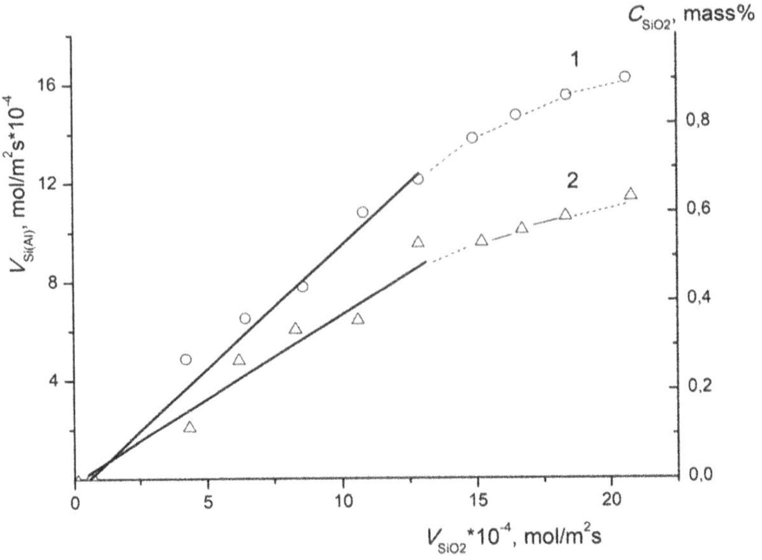

Fig. 4.8 Dependencies of the rate of Si transport into the alloy (1) and content of SiO_2 in cryolite (2) on the rate of SiO_2 input into the electrolyzer

The process proceeds in the diffusion mode and is limited by the delivery of the initial complex designated as Si(IV) to the interface:

$$V_{Si(Al)} = K_m \cdot C_{Si(IV)}, \tag{4.22}$$

where K_m—mass transport constant, m/s; $C_{Si(IV)}$—concentration, mol/m^3.

It was established experimentally (curve 1, Fig. 4.8), that $V_{Si(Al)} = V_{SiO_2}$ (calculated per silicon) at $V_{SiO_2} \leq (14\text{–}15) \times 10^{-4}$ mol/m^2 s. Then

$$C_{Si(IV)} = \frac{V_{SiO_2}}{K_m} \tag{4.23}$$

The total content of silicon compounds in cryolite, which, according to the results of chemical analysis, is formally converted to SiO_2, consists of two parts:

$$C_{Si(cryolite)} = C_{Si(IV)} + C_{Si(II)} = \frac{V_{SiO_2}}{K_m} + \sqrt{K_{in} \cdot C_{Si(IV)} a_{Si(Al)}}$$
$$= \frac{V_{SiO_2}}{K_m} + \sqrt{K_{in}\left(\frac{V_{SiO_2}}{K_m}\right) a_{Si(Al)}} \tag{4.24}$$

or

$$C_{\mathrm{Si(cryolite)}} = A_1 V_{\mathrm{SiO_2}} + A_2 V_{\mathrm{SiO_2}}{}^{1/2}, \tag{4.25}$$

where $A_1 = 1/K_m$; $A_2 = [K_{in} \cdot a_{\mathrm{Si(Al)}} K_m]^{1/2}$.

According to (4.25), one should expect the nonlinearity of graph 2 in Fig. 4.8, which was not actually observed. This is due to the fact that Si is a p-element and for it $K_{in} << 1$ [26, 27], which means that $C_{\mathrm{Si(II)}} << C_{\mathrm{Si(IV)}}$ and, therefore, the second term in (4.25) is negligibly small. It should also be noted that at $V_{\mathrm{SiO_2}} = (14–15) \times 10^{-4}$ mol/m² s the process was still stable, although the product $[C_{\mathrm{SiO_2}}]\{a_{\mathrm{Si(Al)}}\} = (1.6–2.0) \times 10^{-4}$, i.e., slightly exceeded the passivation criterion. An explanation for this was given below.

The performance data of the experimental electrolyzers for three summer months of operation are given in Table 4.2. For comparison, information about the other units of this case is also given.

A decrease in the number of anode effects was noted, which is most likely due to the fact that the current density at which the anode effect occurs in the $Na_3AlF_6–SiO_2$ melt is higher than in the $Na_3AlF_6–Al_2O_3$ melt containing an equiatomic amount of oxygen [28]. No increase in temperature and voltage was observed. Productivity increased by 1.5% and power consumption decreased by 2.9% compared to conventional units. However, the dispenser mechanisms were not reliable and often failed.

Table 4.2 Performance indicators of electrolyzers for the production of Al–Si alloys when the sand was loading with automatic feeding units

No.	Indicators	Experimental electrolyzers		Standard electrolyzers	
		Before tests, 3 months	Accounting period, 3 months	Before tests, 3 months	Accounting period, 3 months
1	Current value (kA)	65.265	64.854	65.265	64.854
2	Specific dosage of sand (kg/day)	–	65–70	–	–
3	Average content of Si in the alloy (mass%)	–	6.5	–	–
4	Specific productivity (kg/day)	455	460	455	453
5	Average temperature (°C)	959.1	959.7	959.1	960.4
6	Voltage drop in the bottom (mV)	330	350	330	340
7	Number of anode effects (day^{-1})	1.72	0.52	1.72	1.63
8	Electric energy consumption (thousand kWh/t)	16.040	15.630	16.040	16.090

Therefore, there were breaks in the loading of sand, which led to a decrease in the average concentration of Si in the alloy in a group of the experimental electrolyzers. On the same units, where the dispensers worked without breakdowns for a long period, alloys with 7.5–8.0 mass% Si were stably obtained.

It was possible to obtain alloys containing 9.5–10.0 mass% Si in the electrolyzers by increasing the sand loading rate to $V_{SiO_2} = 16.5 \times 10^{-4}$ mol/m^2 s. However, it was necessary in this case to maintain a lower level of metal and a higher level of cryolite-oxide melt in comparison with the existing regulations for conventional aluminum electrolyzers. Apparently, the last condition (an increase in the electrolyte volume) ensured stable operation of the unit, since the $[C_{SiO_2}]\{a_{Si(Al)}\}$ value, as in case of producing alloys with 7.5–8.0 mass% Si (see above), was within the same limits, namely (1.8–2.0) $\times$ 10^{-4}. These studies should be considered as preliminary, since they have not been carried out for a long time and only indicate the possibility of technology intensification.

Experiments were carried out on loading disthene-sillimanite concentrate (DSC) using automatic dispensers. Its consumption amounted to ~ 0.28 t/t of alloy, and accordingly, the supply of alumina was reduced. About 50 tons of DSC were processed on four electrolyzers within three months and ~ 167 tons of alloy were produced. No deviations in the operation of the units, in comparison with supplying pure sand, were found.

The works on the improvement of automatic dispensers and their operation were stopped when the reconstruction of the electrolysis department begun. Therefore, the production of Al–Si alloys with the supply of raw materials in a continuous mode was terminated.

4.7.3 *Loading of Kaolin–Alumina Mixture*

The large-scale experiments were carried out using a kaolin–alumina mixture. The installation diagram for its preparation is shown in Fig. 4.9. It is based on the parallel supply of both ingredients with compressed air to the discharge cyclone 2.

The dosage of alumina was carried out by a chamber feeder 3 with a capacity of 3 tons, and kaolin by selecting a calibrated hole 4. The joint supply of ingredients was provided by automation system, which consisted of a pressure sensor 5 located on the alumina pneumatic route and an electromagnetic valve 6 opening compressed air on the kaolin pneumatic route. In total, 38 electrolyzers were used. The raw materials were loaded in a periodic mode 6 times a day after the destruction of the electrolyte crust. The mixture contained ~ 10 mass% kaolin, which should have produced an alloy with ~ 5.0 mass% Si. However, for organizational reasons, kaolin supply was not carried out regularly, which led to a decrease in the average concentration of the alloy.

We compared the operation of experimental and conventional electrolyzers of the same series for two quarters before the start of testing and for the same period during testing. Recorded data are shown in Table 4.3.

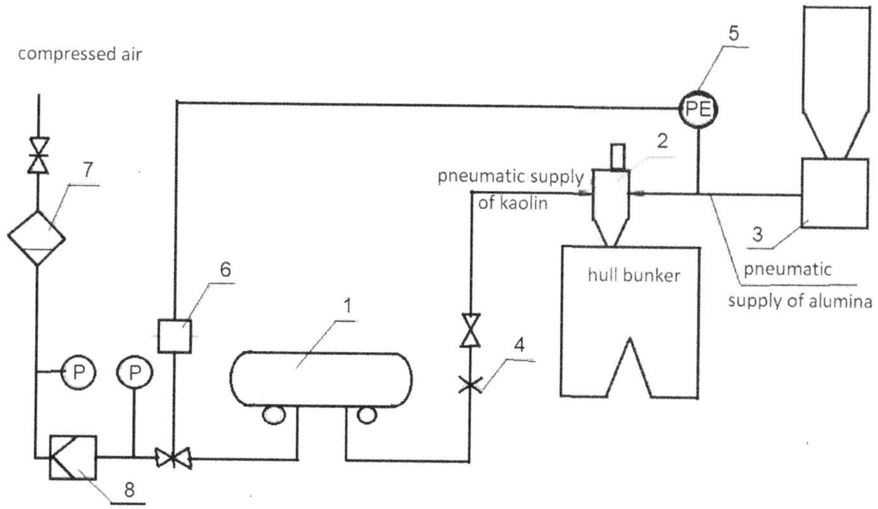

Fig. 4.9 Scheme of the installation for the preparation of kaolin–alumina mixture. Designations: (1) tank with kaolin dust; (2) unloading cyclone; (3) chamber feeder; (4) calibrated hole; (5) pressure sensor; (6) solenoid valve; (7) moisture separator; (8) pressure regulator

Table 4.3 Performance indicators of electrolyzers for the production of Al–Si alloys using kaolin–alumina mixture

No.	Indicators	Period before testing		Accounting period	
		Experimental baths	Standard baths	Experimental baths	Standard baths
1	Current value (kA)	67.780	67.780	67.046	67.046
2	Productivity of the unit (kg/day)	467.75	476.31	470.69	462.70
3	Temperature (°C)	960.7	960.1	961.3	960.9
4	Energy consumption (thousand kWh/t)	16.118	16.094	16.279	16.464
5	Alumina consumption (kg/t)	1968	1968	1830	1969
6	Concentration of silicon (mass%)	–	–	3.7	–
7	Voltage drop in the bottom (mV)	350	340	360	350
8	Number of anode effects (day^{-1})	1.65	1.62	0.68	1.64
9	Anode consumption (g/kAh)	130.0	130.2	135.5	135.3

With a decrease in current value by 1.1%, the productivity of the experimental electrolyzers increased by 0.6%, while for the standard ones it decreased by 2.8%. For the latter, this is explained by the fact that the accounting period fell on the warm season, when there is a decrease in indicators caused by an increase in air humidity and worsening working conditions. The temperature in the experimental electrolyzers increased by 0.6 °C, but less than in standard ones. As in the case of sand loading in continuous mode, there was a significant decrease in the number of anode effects. The energy consumption of experimental electrolyzers turned out to be lower by 1.1% than that of standard electrolyzers, although before the start of the tests this indicator was worse. The consumptions of fluorides and anode mass were the same in both cases.

Such large-scale tests of the technology are highly reliable and therefore allow us to state the following. If the technological process is not disturbed after the switch of aluminum electrolyzers to the mode of producing Al–Si alloys, then their productivity increases (see Chap. 3). In other words, silicon is obtained not instead of aluminum, but beneficently.

When there were no interruptions in the supply of kaolin, the Si content in the alloy reached 5.0–5.5 mass% at a SiO_2 concentration in cryolite from 0.55 to 0.60 mass%. In this case $[C_{SiO_2}]\{a_{Si(Al)}\} = (1.6–1.8) \times 10^{-4}$, which somewhat exceeded the allowable (4.4) (as well as when loading sand using automatic dispensers, see above), but, nevertheless, no violations of the technology were observed. Most likely, this is due to the fact that the value of the passivation criterion (4.4) was obtained in isothermal laboratory cells in the absence of convection. In an industrial electrolytic cell, due to the intensive movement of cryolite melt, the formation of a blocking deposit $[Si(II)]_{solid}$, obviously, occurs at a slightly higher supersaturation of the melt than prescribed by (4.4).

We should make some notes about the peculiarities of the process. They are as follows:

Electrolyzers require close attention of the operating personnel. The technology turned out to be sensitive to fluctuations in the level of cryolite in the electrolyzer. If it is decreased by 10–20% compared to the usual, then sediment is accumulated along

Table 4.4 Silicon balance of the process of electrolytic production of Al–Si alloys for six months

Input of silicon		Output of silicon	
Input article	Mass (t)	Output article	Mass (t)
1. Introduced with kaolin	96.5	1. Removed with alloy	88.3
		2. Remained in the electrolyte	0.2
		3. Unfinished production	6.6
		4. Disbalance	1.4

the sides under the metal. Separate vacuum ladles should be used for pouring the Al–Si alloy. If metal is extracted from conventional electrolyzers using the same ladles, then the residual amount of alloy in them can be a source of aluminum contamination.

Balance of silicon for 6-month period is presented in Table 4.4.

Silicon recovery was 98.6%. The discrepancy of 1.4% is most likely due to the accumulation of Si in the side deposit, rather than its loss in the form of SiF_4. In particular, it follows from a gradual decrease in the Si content in the metal after the kaolin supply was stopped, which did not correspond to the dilution law. These data are in agreement with the results of other researchers, who also recorded no loss of silicon [29–31].

4.8 Physical and Mechanical Characteristics of Electrolytic Alloys

To obtain the comparative data, alloys AK6M2 (GOST 1583) produced by electrolysis (with continuous sand loading) and by fusion (synthesis) were selected. Its composition, wt%: Al: 5.5–6.5; Cu: 1.8–2.3; Mg: 0.30–0.45; Ti: 0.1–0.2 and impurities, mass%: Fe < 0.5; Mn < 0.1; Ni < 0.05; Zn < 0.06. The introduction of dopant additives into the electrolytic alloy was carried out in the mixer of the foundry section.

The study of thin sections on a Nu-2E microscope showed that the electrolytic alloy was modified, in contrast to that obtained by fusion (Fig. 4.10).

The crystallization products of the former are fine-grained and evenly distributed throughout the volume of the alloy, while the synthetic alloy has silicon colonies in

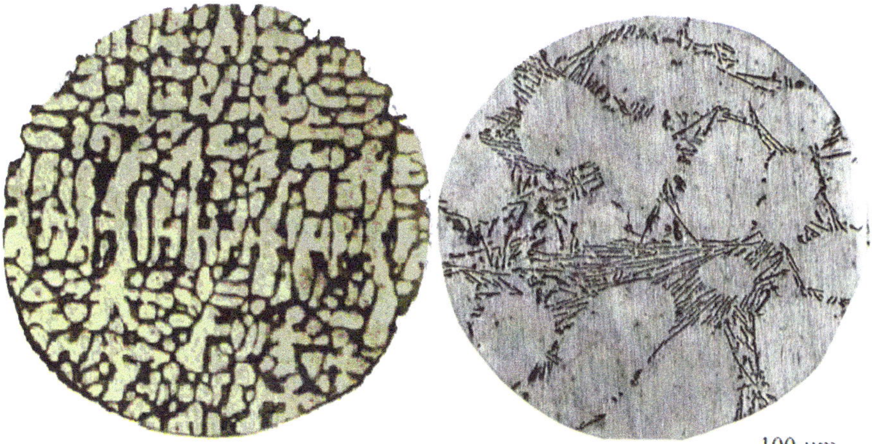

Fig. 4.10 Microstructures of industrial Al–Si alloys grade AK6M2. Left—electrolytic alloy; right—synthetic alloy. Magnification 100

Table 4.5 Mechanical characteristics of AK6M2 alloy, obtained by different methods

Production method	σ (kg/mm^2)	δ (%)	HB (kg/mm^2)
Fusion	13.3 ± 0.3	2.3 ± 0.2	34 ± 3
Electrolysis	16.5 ± 0.4	4.9 ± 0.3	47 ± 4

form of large needles. The light grains of primary precipitates of the α-solid solution of silicon in aluminum are clearly visible. Its microhardness, measured on the PMT-3 device, is 76–82 kg/mm^2. Between the grains of the α-solid solution, along their boundaries, there is an eutectics with a microhardness of 117–120 kg/mm^2.

The electrolytic alloy possesses much better physico-mechanical properties (Table 4.5).[2]

The tests of samples for uniaxial tension showed that the ultimate strength (σ) and relative elongation (δ) of the electrolytic alloy exceed those of the synthetic alloy by 22% and 2.1 times, respectively. Brinell hardness (HB) increased by 38%.

Determination of the hydrogen content in the alloys by the "first bubble" method [32] showed that for the electrolytic alloy this indicator was (0.22 ± 0.03) cm^3/100 g of metal, being at the level of aluminum, while for the synthetic alloy it was (0.30 ± 0.04) cm^3/100 g. No wonder, in the electrolyzer, the production of the alloy occurs at the atomic level under fluoride flux, which ensures its modification, uniform distribution of silicon and low gas saturation. The high quality of the electrolytic Al–Si alloy is confirmed by recent work [33].

4.9 Conclusions

The dissolution rate of alumina in a cryolite melt is slightly higher than that for quartz sand, disthene-sillimanite concentrate, and calcinated kaolin. However, when the process is implemented, the mass of SiO_2-containing minerals introduced into the electrolytic cell is several times less than the mass of alumina, and, consequently, the time of their dissolution will be less than the time of dissolution of alumina.

The mass transfer coefficient of silicon in the electrolyzer for a current of 67 kA turned out to be (0.69 ± 0.10) × 10^{-5} m/s and was an order of magnitude lower than that for aluminum due to the fact that the Si discharge occurs from much more bulky and less mobile nepheline-like groups than aluminum complexes. Both the low mass transfer coefficient of silicon and the limited solubility in the cryolite melt of its intermediate compound Si(II) are factors limiting the production of concentrated alloys.

[2] Dmitriy Pruttskov remembers the following funny story. The VAZ automotive plant in Tolyatti, Russia, was the main customer of casting Al–Si alloys from Zaporizhzhya aluminum plant in times of the USSR. When the machine was broken during the tests of engine pistons made from electrolytic alloy, the officials said: "We do not need such durable material for the pistons. A customer should buy the new pistons after no more than 100,000 km run".

We should also note the work performed twenty years later on the study of the mass transfer of silicon in electrolyzers for a current of 300 kA [34]. Using a similar measurement technique, the authors obtained a much larger value, namely 3.18×10^{-5} m/s. Obviously, this difference is caused by a more intense movement of liquid media in the unit with a larger current value than in our case [11–13]. Having in mind the production of Al–Si alloys in these electrolyzers, we conclude that, under these conditions, a stationary concentration of SiO_2 in cryolite will be established faster. Meanwhile, the possibility of production alloys with higher concentration of silicon requires further investigations, taking into account differences in geometric dimensions (see above).

The technology for obtaining Al–Si alloys in an aluminum electrolyzer is considered as a process taking place in two combined ideal mixing reactors. A mathematical model of the process was stated and solved by the numerical method. The calculations have shown that it is possible to obtain alloys with a Si content, mass%: with a periodic mode of raw material loading 5.0 ± 0.5, and with a continuous mode 7.0 ± 0.5 without violations of the technology. Equations describing the transitional period of operation of the electrolyzer were derived and solved. It has been established that for both regimes of loading raw materials, the time to reach a stationary concentration of SiO_2 in cryolite is ~ 16 h, and the specified concentration of Si in the alloy is achieved in a month.

Using the proposed mathematical model, an analysis of the known results of industrial testing of the method was carried out. It is shown that the destabilization of the process in all cases occurs due to an overdose of SiO_2. As a result, the passivation criterion was exceeded, and sparingly soluble intermediate silicon compounds precipitated.

The main factor of the stability of the technology for producing Al–Si alloys in aluminum electrolyzers is the uniform supply of SiO_2 during the day. Based on the mathematical model for the periodic mode of loading into the cell for a current of 67 kA, the maximum single portion of SiO_2 was calculated, which amounted to $\Delta m = 8$ kg with an interval between loads of 4 h and a round-the-clock implementation of this operation. The fulfillment of these recommendations made it possible to stably obtain the alloys with ~ 5 mass% Si at a SiO_2 content of ~ 0.5 mass% in cryolite.

It has been established that the maximum rate of SiO_2 input into cryolite should not exceed $V_{SiO_2} \leq (14–15) \times 10^{-4}$ mol/m^2 s, or 65–70 kg/day. In this case, alloys with 7.5–8.0 mass% Si were obtained. The technology was used on a group of ten electrolyzers. The productivity of the units increased by 1.5%, and the energy consumption decreased by 2.9% compared to common electrolyzers. There was a decrease in the number of anode effects to 0.52 day^{-1} against the initial 1.72 day^{-1}. No increase in temperature and voltage in this mode of loading of raw materials was observed.

Having in mind the economy of alumina, it is advisable to use alkali-free aluminosilicates as raw materials. The possibility of producing Al–Si alloys by loading a disthene-sillimanite concentrate into an electrolytic cell in a continuous mode is shown. No deviations in the operation of the units compared to feeding them with sand were found.

The most extensive tests were carried out with periodic feeding of 38 electrolyzers with kaolin–alumina mixture. The dosage was set so as to obtain a composition with ~ 10 mass% kaolin, and, accordingly, an alloy with ~ 5.0 mass% Si. However, for organizational reasons, on average, an alloy with a concentration of 70–80% of the specified one was obtained over the entire period of operation. The productivity of electrolyzers increased by 0.6% with a decrease in current value by 1.1%, while in witness electrolyzers it decreased by 2.8%. These results allow us to conclude that when aluminum electrolyzers are switched to the mode of Al–Si alloys production without violations of the technology, their productivity increases. Silicon is obtained not instead of aluminum, but additionally. A decrease in energy consumption by 0.8%, and the number of anode effects down to 0.68 day^{-1}, was recorded. The consumption of fluorides and anode mass did not differ from standard units.

The electrolytic alloy was modified, the products of its crystallization were fine-grained and evenly distributed throughout the volume. The tensile strength and relative elongation of the electrolytic alloy exceed those of the synthetic alloy by 22% and 2.1 times, respectively. The gas saturation was (0.22 ± 0.03) cm^3/100 g and was at the level of pure aluminum, while that of the synthetic alloy was (0.30 ± 0.04) cm^3/100 g.

During the period from 1982 to 1992, using sand, disthene-sillimanite concentrate and kaolin–alumina mixture with periodic and continuous loading of raw materials, 14,570 tons of alloy were produced, including 505 tons of silicon and 2040 tons of alumina were saved. After the collapse of the USSR, the need for Al–Si alloys dropped sharply and, therefore, the operation of the technology was terminated.

References

1. Belyaev A (1944) Metallurgiya legkikh metallov (Metallurgy of light metals). Metallurgizdat, Moscow, 543 p (in Russian)
2. Hannach R, Osborne J, Templeton G, Frazer E, Welch B (1977) Molten salts electrolysis in metal production. In: Proceedings of international symposium, Grenoble, pp 7–13
3. Tabereaux A, McMinn C (1978) TMS light metals 1978, pp 209–222
4. Keller R, Welch B, Tabereau A (1990) TMS light metals 1990, pp 333–340
5. Moxnes B, Gikling H, Kvande H, Rolseth S, Straumsheim K (2003) TMS light metals 2003, pp 329–334
6. Delimarskiy Yu, Pruttskov D, Andriiko A, Chernov R (1983) Ukr Chem J 7:738–742 (in Russian)
7. Pruttskov D, Pirozhkova V, Khvalin A, Chernov R, Delimarskiy Yu (1983) Ukr Chem J 10:1027–1030 (in Russian)
8. Pruttskov D, Kryvoruchko N, Olesov Yu (1994) Ukr Chem J 5/6:433–439 (in Russian)
9. Pruttskov D, Kryvoruchko N, Olesov Yu (1994) Ukr Chem J 7/8:640–646 (in Russian)
10. Pruttskov D (1995) Thesis doctor of science in chemistry. Institute of High Temperature Electrochemistry, Russian Academy of Sciences, 273 p (in Russian)
11. Grjotheim K, Krohn C, Malinovsky M, Matiasovsky K, Thonstad J (1977) Aluminium electrolysis. Aluminium-Verlag, Dusseldorf, 350 p
12. Grjotheim K, Krohn C, Malinovsky M, Matiasovsky K, Thonstad J (1982) Aluminium electrolysis, 2nd edn. Aluminium-Verlag, Dusseldorf, 442 p

13. Thonstad J, Fellner P, Haarberg GM, Hives J, Kvande H, Sterten A (2001) Aluminium electrolysis, 3rd edn. Aluminium-Verlag, Dusseldorf, 359 p
14. Shurygin P, Barmin L, Boronenkov V (1962) Izv VUZov Tsvetn Metall 4:106–112 (in Russian)
15. Schumann W (1972) Stein+Mineralien. B V Verlagsesellschaft, Munchen, 227 s
16. Kachanovskaya I, Sirayev N, Pototskaya N (1969) Tsvetnye Metall (Non-ferrous Met) 9:64–66 (in Russian)
17. Kafarov V (1985) Metodykibernetiki v khimiiikhimicheskoitekhnologii (Methods of cybernetics in chemistry and chemical technology). Moscow, 448 p (in Russian)
18. Frank-Kamenetskii D (2015) Diffusion and heat exchange in chemical kinetics. Princeton University Press, 384 p
19. Stromberg A, Semchenko D (1973) Fizicheskaya khimiya (Physical chemistry). Vysshaya Shkola, Moscow, 480 p (in Russian)
20. Polyakov P, Mozhaev V, Burnakin V, Romanchenko S (1981) Electrochemistry 17:1454–1459 (in Russian)
21. Pruttskov D, Andriiko A, Titayev P (1989) Ukr Chem J 55:569–574 (in Russian)
22. Pruttskov D, Titayev P, Andriiko A, Pirozhkova V (1990) Ukr Chem J 56:470–475 (in Russian)
23. Schaefer S (1974) Rept Invest Bur Mines US Dep Inter 7895:1–15
24. Morachevskiy A, Avaliani A, Mindin V (1978) Zhidkiye kanody (Liquid cathodes). Mziniereba, Tbilisi, 184 p (in Russian)
25. Shakhov G (1948) Metallurgiya (Metallurgy). Metallurgizdat, Moscow, 648 p (in Russian)
26. Smirnov M (1973) Elektrodnye potentsialy v rasplavlennykh khliridakh (Electrode potentials in molten chlorides). Nauka, Moscow, 248 p (in Russian)
27. Baraboshkin A (1976) Elektrokrystallizatsiya metallov iz rasplavlennykh solei (Electrocrystallization of metals from molten salts). Nauka, Moscow, 280 p (in Russian)
28. Belyaev A (1947) Fiziko-khimicheskiye protsessy v elektrolize aliuminiya (Physico-chemical processes in aluminum electrolysis). Metallurgizdat, Moscow, 183 p (in Russian)
29. Reznikov I (1959) Bull Non-ferrous Metall 19:26–28 (in Russian)
30. Begunov A (1967) Bull Non-ferrous Metall 22:36–38 (in Russian)
31. Simko F, Stas M (2009) Can Met Q 48(2):109–113. https://doi.org/10.1179/cmq.2009.48.2.109
32. Hohlov V (1980) Proizvodstvo liteynykh aliumokremniyevykh splsvov (Production of casting aluminum-silicon alloys). Metallurgiya, Moscow, 68 p (in Russian)
33. Wang R, Lu W (2012) Direct electrolytic Al–Si alloys (DEASA), chap 6. INTECH, pp 107–140. https://doi.org/10.5772/52962
34. Al-Mejali JA, Haarberg GM, Bensalah N, Benkahla B-A, Lange HP (2016) TMS light metals 2016, pp 389–394

Chapter 5
Modern Research Trends

5.1 Introduction

We have finished the previous chapter at the time when the USSR collapsed and the successful production of electrolytic Al–Si alloys in the Ukraine was terminated. That was at the beginning of 1990s. Then there was a silence for over 15–20 years. Very few original papers were published on the subject in this period. Some of them were analyzed in the previous chapters.

The interest in cryolite melts containing silica revived at the end of 2000s. The modern research areas can be highlighted as follows:

(1) Development of novel fluoroaluminate electrolytes with lower working temperatures. Such electrolytes, based on potassium cryolite melts, are intended not only (and not mainly) for the preparation of Al–Si alloys, but for the production of master alloys such as Al–Sc and others. The development of inert anodes for such electrolytes is more feasible task than for the classic Hall–Heroult electrolytes due to their lower operation temperature (800–850 °C). This is an additional potential advantage of such melts.

(2) Further investigation of physico-chemical and electrochemical properties of cryolite–silica melts. Again, the main purposes of these studies are the electrolytes for the production of solar-grade silicon rather than Al–Si alloys.

(3) Processing of the spent pot lining. The spent bricks of the lining soaked with fluorine compounds are the serious danger for the environment because of leaching of the fluorides into the soil. So, the processing of these refractory ceramics immediately in the Soderberg electrolyzers is the best option.

(4) Production of oxygen and metal alloys by electrolysis of lunar (or Martian?) soils by means of the electrolysis of cryolite melts. Such possibilities are discussed by NASA scientists and actively studied by Chinese researches.

(5) Use of sodium hexafluorosilicate to obtain Al–Si alloy and cryolite in aluminum electrolyzers. These studies were carried out at the Siberian aluminum plants in 1990s and 2000s.

D. Pruttskov et al., *Electrolytic Production of Al–Si Alloys*,
Monographs in Electrochemistry, https://doi.org/10.1007/978-3-031-29249-1_5

5.2 Novel Potassium Fluoroaluminate Electrolytes

The brazing fluxes for aluminum and its alloys were the first practical applications of the potassium fluoroaluminate–fluorosilicate compositions [1–3].

The compositions of these fluxes were based on the studies of the fusibility and thermal stability of the ternary system K, Al, Si/F melts (Fig. 5.1).

It was found that thermally stable compositions of this system are located inside the triangle KF–K_3SiF_7–K_3AlF_6 (Fig. 5.2), that is, in the left low corner of the diagram in Fig. 5.1. The phase diagram in Fig. 5.2 was studied by visual and DTA methods.

One can see that the diagram in Fig. 5.2 is of simple eutectic type. The composition of the ternary eutectic is (mass%); KF (28)–K_3AlF_6 (10)–K_3SiF_7 (62), with melting

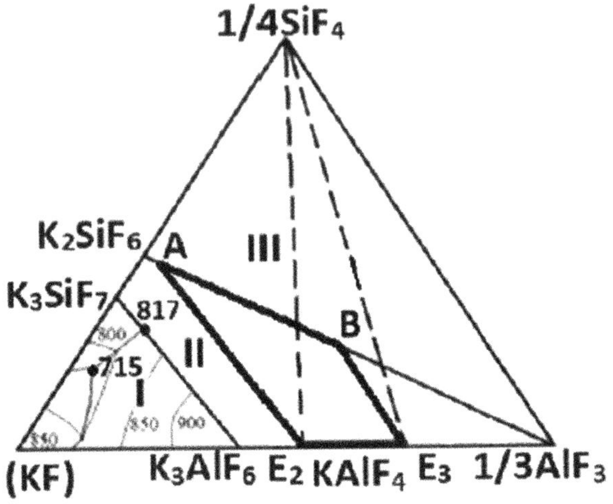

Fig. 5.1 Ternary system K, Al, Si/F

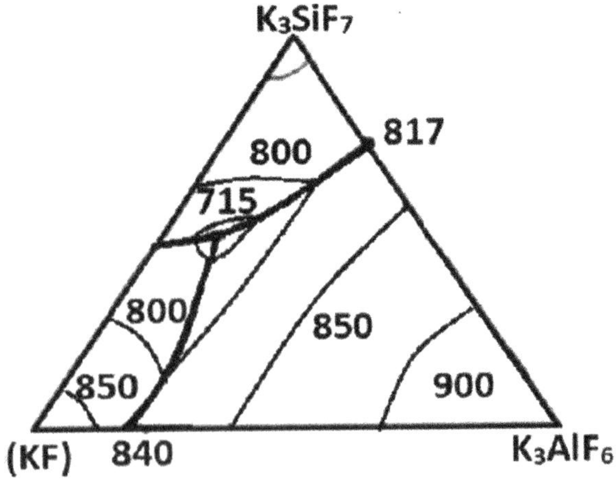

Fig. 5.2 Thermally stable section of the K, Al, Si/F system

point 715 °C. The melting point of the binary eutectics K_3AlF_6 (20 mass%)–K_3SiF_7 is 817 °C.

"Acid" melts outside the triangle of Fig. 5.2, which contains the excess of AlF_3, are not stable. They decompose with the evolution of volatile SiF_4 according to the reaction

$$K_3SiF_7 + AlF_3 = K_3AlF_6 + SiF_4\uparrow \tag{5.1}$$

It was established by measuring the mass losses of compositions inside the area II of Fig. 5.1.

Possibly, the stable melts of KF–K_3SiF_7–K_3AlF_6 system could be used for the production of Al–Si alloys. However, the data on the solubility of silica and alumina as well as electrochemical decomposition of such melts are not yet available.

The electrolyte of KF–AlF_3–Al_2O_3 system for the production of pure aluminum and its master alloys with Si, B, Ti, Zr and Sc was proposed in [4–6]. The operation temperatures of these melts varied in the range from 700 to 800 °C, that is, by 150–250 °C lower than in case of classic sodium cryolite melts used in Hall–Heroult process for many years.

The regularities of joint reduction of Al and alloying element were established and master alloys with several elements (Si, Zr, B, Ti, Sc) were produced in laboratory scale electrolysis of the melts KF–AlF_3–Al_2O_3 at 750–800 °C. Although the results of these experiments were fairly successful, the prospects of industrially scale technology are still uncertain. The main problem, destruction of the graphite bottom of the electrolyzer due to the intercalation of potassium into the graphite, is to be solved for the industrial implementation of these electrolytes. It is clear that the use of electrolytes containing KF requires a fundamental change in the design of the cathode unit. As for the material of the working part of the lining, one can use both exotic substances such as BN and SiC and more affordable products made of corundum concrete with a binder of $CaAl_{12}O_{19}$. The current lead to the liquid cathode can be made of aluminum-resistant TiB_2.

5.3 Physico-Chemical and Electrochemical Properties of Cryolite–Silica System

Several recent works were devoted to the investigation of phase equilibria and chemical reactions in cryolite-based melts containing oxide compounds. Some of them were focused on the corrosion of the cathode refractory lining in aluminum electrolysis cells [7–9].

The products which form by the reactions of aluminosilicate fire clay bricks with cryolite melts belong to the reciprocal quaternary system Al_2O_3–Na_2O–SiO_2–AlF_3–NaF–SiF_4. This system was thermodynamically modeled in [7] using the fitting parameters from some experimental data published in literature. The authors claim that the developed model permits the calculation of thermodynamic or metastable phase equilibria and estimates the behavior of aluminosilicate refractory in-service conditions. It is worthwhile to note that, according to the calculation, the nepheline is formed in the pseudobinary system $NaAlO_2$–SiO_2 at the mole fraction of silica up to 0.6, and the albite is formed at higher content of SiO_2 (above 0.6 mol fraction). It agrees with our data (see Chap. 1).

Some cross sections of the same quaternary system were experimentally studied by DTA in [8]. In particular, the ternary NaF–Na_3AlF_6–$NaAlSiO_4$ system was investigated. The immiscibility area is reported to be adjacent to the NaF–$NaAlSiO_4$ side and is not observed in the binary Na_3AlF_6–$NaAlSiO_4$ system.

The Na_3AlF_6–Al_2O_3–SiO_2 molten system was theoretically studied at high concentrations of silica using the method of the first-principles molecular dynamics simulation [9]. Contrary to the case of low silica content (see Chap. 1), the Si ions with Si–F bonds, such as $[SiF_2O_2]^{2-}$, $[SiFO_3]^{3-}$, $[SiF_3O_2]^{3-}$, were found. We have also admitted the existence of such ions as the unstable intermediates in more dilutes melts.

Some physico-chemical investigations were performed in order to develop the electrolytes for the production of solar-grade silicon [10–13].

The phase diagram of cryolite and sodium metasilicate was studied in [10] by thermal analysis. It contains a small liquation dome between 42.8 and 46.3 mol% of Na_2SiO_3. No volatile substances are formed in this system, as follows from the cryoscopic data, introducing of 1 mol of Na_2SiO_3 into an infinite amount of cryolite will produce 1.5 mol of reaction products.

Also, the density, viscosity and electrical conductivity of the Na_3AlF_6–SiO_2 binary melt were studied by Slovak researchers [11]. Molten mixture Na_3AlF_6–SiO_2 and rapid solidified samples were studied by NMR and XRD methods [12]. Formation of volatile SiF_4 and sodium aluminosilicate $NaAlSi_3O_8$ was reported. Thus, no significant new information was obtained in these recent studies.

Density, conductivity, fusibility, transport numbers and decomposition potentials of cryolite–silica system were measured in [13]. The main target of this Ph.D. study was the development of electrolytic method for preparation of solar-grade silicon (SoG-Si). The following approach was proposed and tested in laboratory scale experiments:

- Electrolysis of Na_3AlF_6–SiO_2 melts with liquid Cu–Si (8 mass%) at 1040 °C in order to increase the content of Si in the alloy up to 24 mass%.
- Slow (0.5 °C/min) cooling of the obtained alloy from 1250 to 780 °C. This is the so-called solvent refining. As a result, silicon dendrites are formed and embedded into the copper matrix.
- Crushing the solid alloy to form particles of about 150 μm size.
- Dissolution of copper matrix by nitric acid and/or aqua regia. The author denotes this final operation as "acid leaching".

Thus obtained, the silicon was far from being of SoG-Si purity. For example, it contained large amount of copper (1426 ppm), which was unavoidable because of certain solubility of Cu in solid Si. The contents of Fe, P and Mo were also relatively high. Taking into account the purity problems as well as the ecological issues (a large amount of harmful nitrogen oxides are evolved during the acid treatment), we do not believe in the possibility of commercialization of this process.

Once again, we would like to emphasize that the Na_3AlF_6–SiO_2 binary system is not thermally stable, since SiF_4 is separated into the gas phase from the melt. Figures 1.5 and 1.6 of Chap. 1 illustrate this fact. The vapor pressure above this melt is more than an order of magnitude higher than that for individual cryolite, and SiF_4 volatilization is observed for more than an hour. Therefore, the composition of the melt changes during measurements and it is not clear to what content of SiO_2 the obtained values refer.

We should also note that cryolite melts are not a good solvent for obtaining high-purity silicon. They have a fairly high melting point. In addition, aluminum will also be discharged at high current densities. The low-melting systems based on K_2SiF_6 are preferable. They have already been successfully used for this purpose (see [14] and references therein).

5.4 Processing of Spent Pot Lining

The papers [7–9] considered in the previous section were focused on the corrosion of pot lining ceramics and prolongation of its life span. Further on, we will consider the recent works on the processing of used pot lining after the pot shutdown.

Figure 5.3 represents the pot lining of aluminum electrolyzer. It serves for 2000–3000 days, then deteriorates and must be replaced. The spent pot lining (SPL) must be utilized in a most safe and efficient way.

As follows from Fig. 5.3, SPL consists of two fractions—carbon from the old cathode blocks, and ceramic from the old refractory and insulating bricks. Only the second fraction can potentially be used as raw material for the production of Al–Si alloys in the industrial electrolyzers.

As for the carbon fraction, it was proposed to use it as the reducing agent in the carbothermic process of Ferro Silico Manganese alloys smelting [15].

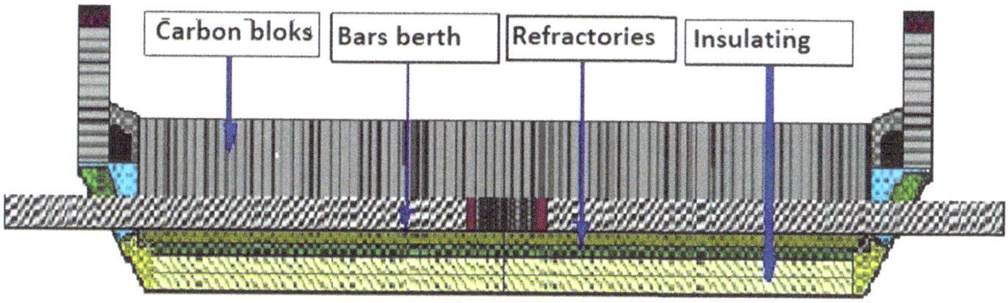

Fig. 5.3 Typical scheme of pot lining

Testing of crushed ceramic fraction of SPL as the raw materials for the production of Al–Si alloys was described in [16]. Two series of experiments have been performed. First, the laboratory scale when only SPL was used and second, industrial tests in 115 kA electrolyzers which were fed with the mixture of SPL and pure alumina. It was found that the dissolution of SPL material into the electrolyte is slower than for alumina, "which requires some special care in the way that the SPL material is added to the cells, i.e., large and concentrated batches should be avoided". It was reported that no complications occurred when the industrial cell was fed so that the content of SPL would not exceed about 15 mass% of that of normal alumina. Otherwise, problems with "sludging" were encountered. About 60 days after the SPL feeding had started the bath temperature rose to about 1000 °C and the cell current was reduced by connecting a resistor in parallel (shunting the cell) for 10 days.

Apart from electrolysis, other methods of SPL utilization have been proposed, for example, the use in cement production technology [17, 18]. However, the electrolytic method of processing in industrial electrolyzers is much better both from an economic and environmental point of view. It permits to dispose the production wastes directly at the place of their formation/which is a significant advantage.

The main condition for the success of this process is the correct feeding of the electrolyzer to avoid the passivation phenomena and the related technological complications (see Chaps. 2 and 4).

5.5 Can a Human Breathe on the Moon?

The answer is—***not yet***. However, it might be possible in future, when/if the projects for the production of oxygen and metal alloys by the electrolysis of lunar regolith in molten cryolite would succeed.

The supply of oxygen for space exploration to sustain life and/or to be used as a component of rocket fuel [19] is a challenging problem.

The crusts of extraterrestrial objects, like Moon, Mars, asteroids, are inexhaustible source of oxygen. It exists there in form of metal oxide compounds, which must be decomposed to release oxygen. Evidently, it could be done only by electrolysis producing metal at the cathode and oxygen at the anode.

Electrolytic production of oxygen and metal from lunar regolith in $CaCl_2$–CaO molten electrolyte at 850 °C was considered in [20, 21]. Solid solution of $CaRuO_3$–$CaTiO_3$ was suggested as the inert anode material for this process.

It was announced that China planned a crewed lunar landing mission in the 2030s, and possibly building a base near the lunar south pole with international cooperation (https://en.wikipedia.org/wiki/Chinese_Lunar_Exploration_Program).

Thus, the China scientists are active in this research field. Their efforts are focused on the development of oxygen and metal production technologies based on the electrolysis of sodium cryolite melts. The laboratory experiments were carried out using the simulants of lunar soil, such as volcanic rock [22, 23] or coal fly ash [24], with chemical composition close to the samples delivered by Apollo missions.

An electrochemical method of preparation of Al–Si master alloy was investigated in cryolite melt adding 5 mass% volcanic rock at 1233 K [22]. The cathodic product obtained by galvanostatic electrolysis contained in mass%: Al—90.5, Si—4.4, Fe—1.9 and Ti—0.2. No data on current efficiency were reported.

Similar experiments with the same rock were described in [23]. The composition of anode (Fe: 56, Ni: 44 mass%) gases was studied by chromatographic method. During the electrolysis, oxygen concentration increased and reached 97.1% after 1 h. The cathode product consisted of (in mass%): Al: 56.70, Si: 40.80, Fe: 2.25 and Ti: 0.25. Again, no data on current efficiency were reported either for cathode and anode process. It should be pointed out that such concentrations of alloys are not achievable under industrial electrolysis conditions due to the formation of a poorly soluble compound $[Si(II)]_{solid}$ (see Chap. 4).

An approach that combined aluminothermic reduction and electrolysis process in cryolite molten salts to produce both Al-based alloy and oxygen from lunar soil simulants was proposed in [24].

First, the coal fly ash or other simulant is dissolving in molten cryolite. After the dissolution, aluminum is added to the melt and aluminothermic reduction takes place. Aluminum-based master alloy forms, which is separated from the salt phase (alumina-rich cryolite) and the latter is subjected to electrolysis with inert anode to produce pure aluminum and oxygen. According to the laboratory tests [24], the alloy containing (in mass%) Si: 7.5, Ti: 3.2 and Fe: 1.0 is formed at the first stage.

The electrolysis of alumina-rich cryolite melts at 1233 K using Fe–Ni alloy inert anode allows to produce pure Al and oxygen. It was concluded that one "can extract metals and oxygen by lunar resource in-situ utilization, which will provide a significant alternative for lunar exploration". Nevertheless, the content of impurities in the resulting cryolite-alumina melt is not reported. Therefore, the production of high-purity aluminum in the second stage is questionable.

The central point of the possible scheme (Fig. 5.4) is the aluminothermic reduction of the lunar rock dissolved in the cryolite melt. Therefore, we should consider this process in more detail than it was done in Chap. 2. Our studies of the kinetics of this reaction (Fig. 2.4) were carried out in corundum crucibles while other researchers used graphite crucibles for this purpose [24–26]. This is important for reliable data to be obtained. The differences in physico-chemical mechanisms are demonstrated by the thermogravimetric study of this reaction [27].

The experimental conditions were as follows. 300 g of cryolite melt (c.r. = 1.7) with 2 mass% SiO_2 was placed in a corundum crucible with a capacity of 0.2 l. Smaller corundum or glassy carbon crucibles (0.02 l) containing 10 g of aluminum were attached to the device for automatic weight registration with a platinum thread. The smaller crucible was first placed over the surface of the cryolite and was immersed after the melting of aluminum. Several stages of interaction are observed (Fig. 5.5).

The mass of both samples first increases (section OA). It can be explained by the formation of a passivating deposit on the aluminum surface (see Chap. 2). Then the mass stabilizes (section AB), which indicates the inhibition of the reaction. In the case of a glassy carbon crucible, a sharp increase in mass was observed after ~ 30 min (section BC). As it turned out, this was caused by the deposition of a precipitate on the

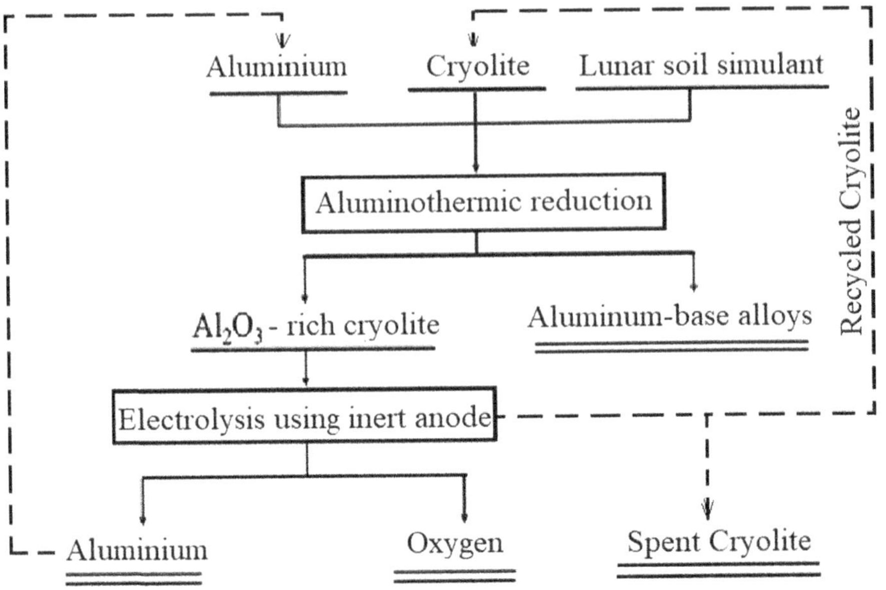

Fig. 5.4 Possible scheme of metal and oxygen extraction from lunar soil simulant by aluminothermic reduction followed by molten salt electrolysis. See explanations in the text

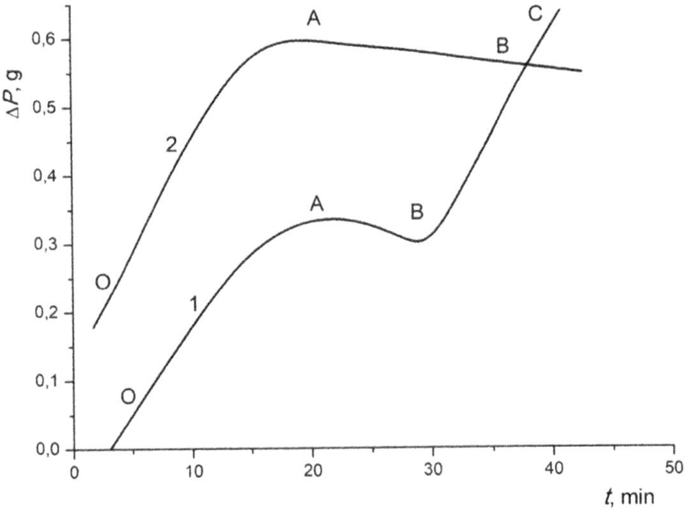

Fig. 5.5 Gravimetric curves of weight change of crucibles with aluminum immersed in cryolite melt. Designations: (1) glassy carbon; (2) corundum crucible. See explanations in text

free surface of the crucible contacting with the melt. X-ray phase analysis revealed that it contains silicon. The formation of a short-circuited electrochemical galvanic cell could be the cause of this phenomenon

$$CG|\text{cryolite}-Al_2O_3-SiO_2|Al \tag{5.2}$$

Anodic dissolution of aluminum and cathodic deposition of silicon take place in this cell. For verification, the following galvanic circuit was assembled:

$$Al|\text{cryolite}-Al_2O_3||\text{cryolite}-Al_2O_3-SiO_2|CG \tag{5.3}$$

The scheme of this circuit is similar to that shown in Fig. 2.1 with the difference that the outer crucible was made of glassy carbon and did not contain aluminum. The current lead was attached directly to the crucible. The open-circuit voltage of this cell was equal to ~ 1.7 V. When the electrodes were connected to the external circuit, a current of ~ 0.3 A passed steadily for more than 3 h, the cathode area being ~ 50 cm^2 and the anode area ~ 4 cm^2. Silicon was also found on the working surface of glassy carbon, which confirmed our hypothesis.

It is clear that carbon crucibles cannot be used because of the effects considered above. On the other hand, it follows that carbon pot lining of the industrial cells containing silica in the cryolite melts cannot be recommended for aluminothermic reduction process due to the same reasons.

Let us take a closer look at the scheme of material flows of the technology for processing lunar rock into aluminum alloy and oxygen. The chemical composition of the lunar rock (according to [22, 23]) is within the range (in mass%): Al_2O_3: 15.0–17.4, SiO_2: 44.9–52.7, TiO_2: 1.6–2.9, FeO: 10.4–13.1, MnO: 0.14–0.34, P_2O_5: 0.14–0.66, Na_2O: 0.7–3.8, K_2O: 0.5–3.0, MgO: 4.4–9.4, CaO: 9.4–10.7, LOI: 0.1–0.5. The first six oxides are formed by more noble elements and will be recovered from the cryolite melt, while oxides of alkali metals, magnesium and calcium will accumulate in the latter in the form of fluorides. The calculated average composition of the resulting alloy will be as follows, mass%: Al: 24.2, Si: 51.2, Ti: 3.0, Fe: 20.7, Mn: 0.4, P: 0.5. The melting point of this alloy will be higher than that of cryolite, and when trying the direct electrolysis of lunar soil, a deposit in form of a metal-salt "rod" will form on a vertical cathode. The growth mechanism of such deposits in the melts containing the compounds of polyvalent metals is considered in detail in [28]. Processing of the metal-salt "rod" is usually carried out by means of a multi-stage hydrochemical technology, which, for obvious reasons, cannot be implemented under the conditions of a lunar mission. In other words, the cathode product cannot be used and will be wasted in this case.

Electrolysis on a liquid aluminum cathode is more preferable, since complete separation of the metal from the fluoride-oxide melt is ensured. However, due to the effect of passivation phenomena, the maximum achievable silicon content in the alloy is limited to ~ 10 mass% or less (see Chaps. 2 and 4). Then, in addition to the lunar rock, alumina should be also loaded into the electrolyzer. Obviously, the latter (as well as cryolite) must be delivered from Earth. As an option, the scheme presented in Fig. 5.4 can be applied when aluminum is delivered to the Moon.

As follows from the practical experience of the authors, the aluminothermic reaction should be carried out in electric furnaces with intensive stirring (e.g., equipped with an inductor [29]) loading lunar rock in a continuous mode, because we know that the attempts to obtain more concentrated alloys will result in the accumulation of a large amount of $[Si(II)]_{solid}$ precipitate (see Chaps. 2 and 4). Although the aluminum obtained by electrolysis will be contaminated with impurities, it can be used for an aluminothermic reaction. Furthermore, it is necessary to ensure a stable content of alumina in the cryolite melt, because a sharp decrease in Al_2O_3 concentration leads to the oxidation of fluorine ions, which brings about dissolution of the anode due to the formation of iron and nickel fluorides. This harmful phenomenon can be eliminated by constant recirculation of the cryolite-oxide melt.

The resulting Al–Si alloy will be contaminated with Fe, Ti, Mn and P impurities, but it can be processed in separate metallurgical plants. It should also be noted that impurities of more active metals (Na, K, Mg and Ca) will accumulate in the cryolite-oxide melt, and thus its periodic replacement be will necessary.

Taking into account the above reasoning, the diagram in Fig. 5.4 was supplemented with recirculation flows, which are indicated by dashed lines. Since the gravity on the Moon is much less than on the Earth, the delivery of the Al–Si alloy and spent cryolite in the opposite direction will be reasonable.

An inevitable disadvantage of this technology should be pointed out. As a matter of fact, the Al–Si alloy obtained by aluminothermy contains < 10 mass% of silicon (see above). This amount of Si, as well as the reduced Ti, Fe and Mn, actually correspond to the chemical equivalent of oxygen obtained in the subsequent electrolysis. Whence it follows that it is necessary to deliver ~ 9 times more aluminum to the Moon than is required by the stoichiometric equation for the decomposition of its oxide. With such an organization of the process, the overall energy yield of oxygen (in g/kWh) is negligible, and the logistics costs are enormous. Therefore, it can be argued a priori that it will be cheaper to transport compressed oxygen to the Moon. However, there is a way out of this dead end. More efficient is the technology shown in Fig. 5.6.

In our opinion, direct electrolysis of the lunar soil should be carried out to obtain a metal-salt rod (see above) and oxygen at the first stage. The design of the electrolyzer will be the same as for the production of zirconium (Fig. 5.7) [30].

The unit is a sealed steel container. Its lower part is forcibly cooled, which ensures the formation of an artificial garnissage. The curvilinear-shaped cathode periodically rises from the melt, and the resulting product is cut off from it. The pieces of cathode product enter the collection tank through the lock device. It was possible to create an industrial unit for a current of 40 kA[1] by increasing the number and area of

[1] Data for 1991, when Dmitriy Pruttskov worked in collaboration with Institute of Rare Metals (Moscow, USSR). Similar cell for titanium production was built at the Pilot Plant of the Titanium Institute (Zaporizhzhia, Ukraine) in the early 1990s. However, due to the collapse of the USSR, funding for the work and its operation was stopped and it was dismantled for scrap.

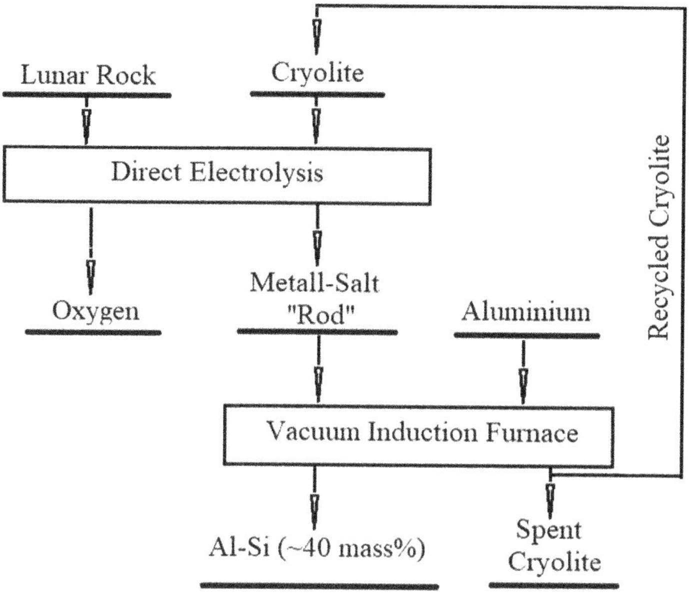

Fig. 5.6 An alternative scheme for obtaining oxygen and Al–Si alloy from lunar soil

electrodes. Electrolyzers of this design are successfully operated at OJSC Chepetsky Mechanical Plant (Glazov, Russia).

The metal-salt "rod" may contain > 30 mass% of inorganic compounds [14, 28]. As we have said, its processing in the conditions of lunar mission is not possible. The obvious solution is the immersion of this "rod" into the intensively stirred molten aluminum. In this case, the total content of silicon and other dopants in the alloy is limited only by the temperature in the furnace. For example, at 1000 °C the composition of the alloy is Al–Si (40 mass%), which corresponds to the liquidus line of the phase diagram. This value is much higher than can be achieved by aluminothermal process. Thus, the amount of aluminum to be transported from the Earth can be reduced significantly.

A further increase in temperature to obtain more concentrated alloys should result in the intense volatilization of fluorides and thus is not advisable. Since the cathode product, in addition to silicon, contains Ti, Fe and Mn, the designation of the alloy in Fig. 5.7 is conditional. The resulting cryolite is returned to the cycle and, as impurities accumulate, is removed from the process. Additional studies are needed to compile an accurate material balance.

There is an alternative option in the scheme. The pieces of metal-salt "rod" can be crushed on the Moon, packaged and returned to Earth, where it can be remelted with aluminum. Then, instead of aluminum, it is necessary to deliver additional amounts of cryolite to the Moon. The differences in the bulk weights of these powders and the specific densities of aluminum and Al–Si alloys should be considered when calculating transport flows in this case.

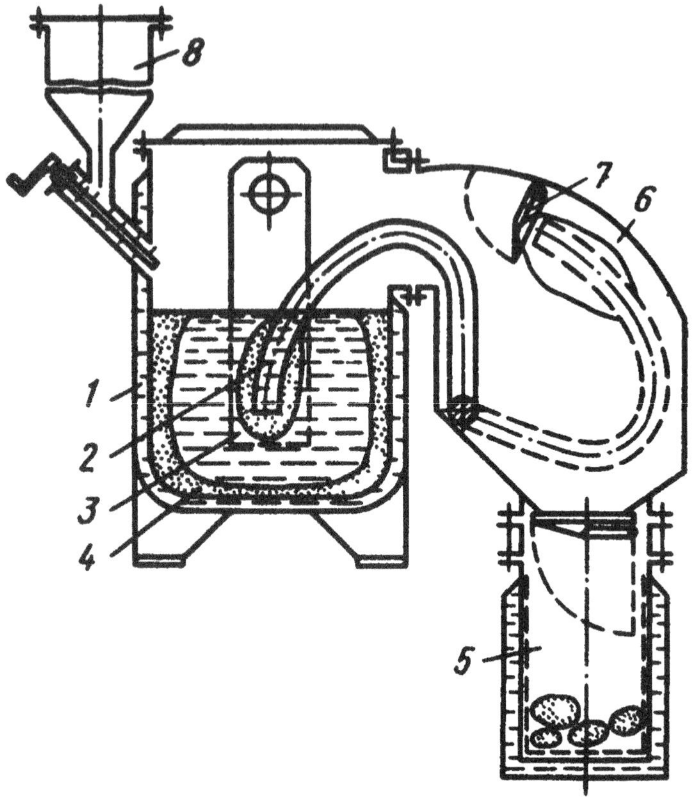

Fig. 5.7 Scheme of a hermetically sealed continuously operated electrolyzer with an artificial garnissage according to [30]. Designations: (1) cooled case; (2) cathode; (3) anode; (4) garnissage; (5) product container; (6) deposit cutting chamber; (7) knife; (8) hopper with raw materials

The choice of the flow diagram of final process requires a thorough feasibility study, taking into account the number and type of equipment involved and logistical costs.

It is clear that the development of reliable and efficient **inert anodes** is very important for the production of oxygen on the space missions and for the reduction of greenhouse gas emissions on Earth as well. The state of the art in this field was reviewed recently in [31].

Three types of inert anodes are considered: (1) Ceramics (individual oxides of Ni, Sn, Fe, Cu and their combinations); (2) Cermets, i.e., sintered mixtures of ceramics with metals; and (3) Metals (alloys). The ceramic anodes ($NiFe_2O_4$, SnO_2 and others) were the subject of early studies. Their disadvantages are low conductivity and weak mechanical properties. That is why most of the modern researches are focused on metals and, to the less extent, on cermets.

The Fe–Ni alloys are considered to be the most promising [32, 33]. They are fairly stable in low-temperature potassium cryolite melts (see above) at 700–800 °C and slightly less stable in classic sodium cryolite electrolytes at 950 °C. The large-scale experiments on their use were carried out at the RUSAL scientific and technical center (Krasnoyarsk) [34]. The corrosion stability is ensured by the formation of protective $NiFe_2O_4$ spinel layer. However, slow corrosion takes place which results in the contamination of aluminum with Fe and Ni.

However, the Fe–Ni anodes might well be used in the electrolysis of lunar regolith where the impurity concerns are not important. Turning back to the lunar program, we would like to note the following. In this case, it would be quite reasonable to use low-melting electrolytes based on the Li, Na, K/F system. This is important not only for the service life of the Fe–Ni anode. The fact is that lowering the temperature of the electrolysis process should ensure a decrease in the saturated vapor pressure over the Al–Si/cryolite system, and, accordingly, the amount of harmful emissions consisting of M, AlF and $MAlF_4$ (see Chap. 3). The utilization of these vapors in the conditions of the closed space of the Lunar Base will require the development of aspiration systems undoubtedly more complex than those at the Earth's aluminum plants. Obviously, highly qualified metallurgical engineers with experience in the design, construction and operation of relevant plants should be involved in this project.

5.6 Use of Na_2SiF_6 for the Production of Al–Si Alloys and Cryolite in Aluminum Electrolyzer

Sodium hexafluorosilicate, as a by-product, is formed in large amounts at chemical plants producing phosphorous fertilizers from fluorapatite as raw material. Therefore, the problem of its effective utilization is urgent.

It is not possible to introduce it into the aluminum electrolyzer directly, since it decomposes with the evolution of SiF_4 into the gas phase at a temperature above 700 °C. Though, it is known that sodium silicofluoride easily reacts with alumina to form fluoroaluminates (Na_3AlF_6 and $Na_5Al_3F_{14}$) and silica in the temperature range from 550 to 700 °C [35]. Thus, it is possible to obtain raw materials for the production of Al–Si alloys by pre-preparing a mixture of these ingredients and heat treatment and additionally to save fluoride salts. It was proposed to load the mixture directly onto the electrolyte crust, where the required temperatures are established to eliminate an additional operation of preliminary heat treatment [36, 37]. Consequently, the process of electrolysis of the melt actually proceeds in Na_3AlF_6–Al_2O_3–SiO_2, which regularities have been considered in detail (see Chaps. 2 and 4). The fluoroaluminate accumulates in the unit and is periodically merged and transferred to other baths.

The first experiments were carried out in an electrolyzer for a current of 132 kA. The electrolyzer worked stably with silicon content in the metal about 4.8 mass%, and SiO_2 in the electrolyte 0.49 mass%. In this case, the passivation criterion was approximately $[C_{SiO_2}]\{a_{Si(Al)}\} = 1.15 \times 10^{-4}$, which is in agreement with our data (see Chaps. 2 and 4). Due to the accumulation of the fluoroaluminate in the bath, the volume of the electrolyte increased by 20–50% compared to the usual one, which results in the oxygen depletion and more frequent anode effects than in the case of using oxide raw materials.

Subsequently, the scale of experiments was widened. The following number of electrolyzers was used in different plants: for a current of 132 kA: 6 units; for a current of 155 kA: 3 units; for a current of 175 kA: 1 unit.

Balance measurements were carried out during a quarter of year on the three units for a current of 155 kA. The composition of the mixture was as follows, mass%: Na_2SiF_6: 60 and Al_2O_3: 40. The initial dosage of the mixture was (in average) ~ 500 kg/day, and then increased to ~ 1200 kg/day. During the testing period, the electrolyte composition was not corrected with fluoride salts (except for CaF_2), while the cryolite ratio was self-established at a level of 2.6. The silicon concentration in cryolite was in the range of 0.15–0.6 mass% (a larger value, obviously, just after loading) and averaged to 0.31 mass%. During the balance tests, 53.65 tons of cryolite melt were obtained. Of these, 17.45 tons were disposed in the experimental baths themselves (58 kg/t) and 36.2 tons were used for liquid feeding of other electrolyzers.

The extraction of silicon into the alloy was 70.5%, which the authors explain both by the loss of SiF_4 during the thermal treatment of the mixture and by the accumulation of SiO_2 in the accretions of the electrolyzer. The latter is confirmed by a gradual decrease in the silicon content in the metal after the loading the mixture into the electrolyzer was stopped. The regularity of this decrease did not correspond to the theoretically predicted. Taking this into account, the recovery increases by ~ 4%. We have also observed similar phenomenon (see Chap. 4).

A slight increase in temperature (in average by 5 °C) was noted, which, in our opinion, is mainly due to an increase in the number of anode effects. No other violations, including the formation of sediments on the bottom, were observed. The most important result is the achieved increase in the productivity of electrolyzers. Thus, the average daily production was 1093 kg against 1055 kg in conventional baths. That is, if the organization of the technological process is correct and technology violations are avoided, silicon is produced additionally and not instead of aluminum, with virtually no energy costs. The data of these authors are an independent confirmation of the reliability of our laboratory measurements and industrial tests presented in Chaps. 3 and 4.

The disadvantage of the technology is significant losses of silicon in the form of SiF_4, which requires additional means to capture and neutralize it.

5.7 Conclusions

Thus, the works of recent years discussed in this chapter confirm our earlier conclusions about the stability and chemical reactions in Na_3AlF_6–Al_2O_3–SiO_2 melts, as well as about the causes of technological violations and ways to eliminate them. Essentially, no new and significant results have been obtained.

However, most of the works substantiate the feasibility of producing alloys with Si, Ti, Mn and Sc in existing aluminum electrolyzers by loading the corresponding oxides [38]. It is worthwhile to remind that extensive industrial tests were carried out at the Zaporizhzhia aluminum plant (Ukraine) in the 1930s [39], and later in the 1970s–1990s, to produce alloys in Soderberg electrolyzers for a current of 65 kA not only with silicon, but also with titanium and zirconium [40, 41]. It was possible to obtain alloys containing ~ 1.5 mass% without complications by loading of TiO_2 and ZrO_2 onto the electrolyte crust. The attempts to increase the concentration of the alloy failed. Similar violations were observed, namely the formation of sediment at the bottom, an increase of the cell voltage and temperature.

It was also found that the actual accumulation of alloying metals in the initial period was lower than the calculated one. An increase of the mass of a single portion of oxide results in a significant decrease of the accumulation rate. Also, after the input of raw materials is stopped, the content of alloying elements in aluminum decreases very slowly compared to the dilution law. The authors explain these peculiarities by the accumulation of a part of the alloying metal oxide into the side garnissage and precipitate with slow subsequent reduction after the dosage of oxides is stopped. However, this explanation disagrees with the later work of the same authors [42] where they have measured the dissolution rate of TiO_2 and ZrO_2 in a cryolite melt. According to the data obtained, this value is close to that for alumina. Since the mass of their one-time portion was approximately 0.01 of the mass of the loaded alumina [40, 41], which was much less than the solubility limits [43], it is difficult to admit a significant accumulation of the oxide in the precipitate.

In our opinion, it is a poorly soluble compound of the alloying element in an intermediate oxidation state that forms the passivating deposit rather than the initial oxide. This conclusion is also supported by the fact that a passivating precipitate was found at the Al/cryolite–oxide interface (see Chap. 2). We have already considered the conditions for the formation of a poorly soluble compound and its effect on the technology using the production of Al–Si alloy as an example (see Chaps. 2 and 4). Thus, we can make some general remarks on the organization of the process of producing other alloys in industrial electrolyzers.

The main condition for the stable operation of electrolyzers is the regular loading of alloying metal oxides throughout the day. If these oxides are loaded periodically on the electrolyte crust, maximum allowable single dose of the oxide (Δm) is the main parameter. The exact calculation of this parameter requires detailed physico-chemical and technological studies. Anyway, to begin with, we recommend a small dose. The increase of the dose is recommended not earlier than a month later, because this time is necessary to achieve the stationary concentration of metal in the cathode product (see Chap. 4). The main determining parameter is the specific loading rate (mol/m^2 s) if the feeding is carried out continuously by automatic dispensers. Again, we recommend starting the feeding with low download speed, increasing it no sooner than after a month operation. Naturally, the technology parameters will be different for each particular alloy.

A new research field has emerged—the electrolysis of lunar soil to produce an Al–Si alloy and oxygen—which we were not able to foresee in the 1980s when most of our data have been obtained. This lunar raw material is polymetallic, which, in addition to SiO_2, contains large amount of FeO and TiO_2. Obviously, the described passivation phenomena would be observed in course of its processing, which means that the technological methods should be the same. We are confident that this book will be useful for the development of research in this area also.

References

1. Andriiko AA, Panov EV, Yakovlev BV, Khorunov VF, Sabadash OM (1997) Ukr Chem J 63:121–124 (in Russian)
2. Khorunov VF, Sabadash OM, Andriiko AA (1998) Investigation of fusibility and chemical interaction in the K, Al, Si/F salt system fluxes for high-temperature brazing of aluminium. In: Proceedings of international conference on brazing, high temperature brazing and diffusion welding, Aachen, 16–18 July 1998, pp 192–201
3. Khorunov VF, Sabadash OM (2009) Weld Mater Test 4:46–50
4. Suzdaltsev AV (2022) Elektrodnye protsessy pri poluchenii aliuminiya i ego ligatur v rasplave systemy KF–AlF_3–Al_2O_3 (Electrode processes in the production of aluminum and its master alloys in the melts of KF–AlF_3–Al_2O_3 system). Dr of science thesis in electrochemistry, Institute of High Temperature Electrochemistry of Ural Division of Russian Academy of Sciences, Ekaterinburg, 259 p (in Russian)
5. Pershin P, Suzdaltsev Yu, Zaikov YuP (2016) J Electrochem Soc 163:167–170
6. Suzdaltsev AV, Nikolaev AYu, Zaikov YuP (2021) J Electrochem Soc 168(4):046521
7. Lambotte G, Chartrand P (2013) TMS light metals 2013, pp 911–916
8. Rutlin J, Grande T (1999) J Am Ceram Soc 82:2538–2544
9. Feng Y, Li M, Hou W, Cheng B, Wang J, Li H (2021) ACS Omega 6:3745–3751
10. Simko F, Boca M (2007) Helv Chim Acta 90:1529–1537
11. Korenko M, Vaskova Z, Priscak J, Simko F, Ambrova M, Shi Z (2015) Silicon 7(3):261–269
12. Šimko F, Rakhmatullin A, Korenko M, Bessada C (2021) J Mol Liq 328:115453
13. Sokhanvaran S (2014) Molten salt electrodeposition of silicon in Cu–Si. Ph.D. thesis, University of Toronto, 142 p
14. Panov E, Andriiko A, Boiko O, Yakovlev B, Borovyk O (1997) Ukr Chem J 63:7–11 (in Russian)
15. Kruger P (2011) TMS light metals 2011, pp 275–280
16. Moxnes B, Gikling H, Kvande H, Rolseth S, Straumsheim K (2003) TMS light metals 2003, pp 329–333
17. Birry L, Poirier S (2020) TMS light metals 2020, pp 1237–1242
18. Jawi M, Chow C, Pujari S, Pan M, Kulkarni T, Mahmoud M, Akasha H, Abdulla S (2020) TMS light metals 2020, pp 1251–1260
19. Sanderson K (2009) How to breathe on the moon. Nature 2009. https://doi.org/10.1038/news.2009.803
20. Juzeliūnas E, Fray DJ (2020) Chem Rev 120(3):1690–1709
21. Schwandt C, Hamilton JA, Fray DJ, Crawford IA (2012) Planet Space Sci 74:49–56
22. Liu A, Shi Z, Xu J, Xu X, Gao B, Wang Z (2016) JOM. https://doi.org/10.1007/s11837-016-1898-x2016
23. Liu A, Shi Z, Xu X, Gao B, Wang Z (2017) J Electrochem Soc 164:H126–H133
24. Liu A, Shi Z, Hu X, Kubikova B, Boca M, Gao B, Wang Z (2017) J Alloys Compd 718:279–287
25. Qiu Z, Zhong L, Grjotheim K, Kvande H (1987) Aluminium 63(12):1247–1250
26. Gorji YM, Soltanieh M, Habibolahzaden A (2007) Can Metall Q 46(4):385–390
27. Delimarskiy YuK, Pruttskov DV, Andriiko AA, Chernov RV (1983) Ukr Chem J 49:739–742 (in Russian)
28. Andriiko AA, Andriiko YuA, Nauer GE (2013) Many electron electrochemical process, vol XIX. Springer, 167 p

29. Pruttskov DV, Kotsur VS, Lysenko VG, Kontsur EP, Khvalin AP (1987) Izv VUZov Non-ferrous Metall 2:32–36 (in Russian)
30. Baryshnikov NV, Geger VE, Denisova ND, Kazain AA, Kozhemyakin VA, Nekhamkin LG, Rodyakin VV, Tsylov YuA (1979) Metallurgiya tsirkoniyai gafniya (Metallurgy of zirconium and hafnium). Metallurgy Publishing House, Moscow, 208 p
31. Padamata SK, Yasinskiy AS, Polyakov PV (2018) Progress of inert anodes in aluminum industry: review. J Sib Fed Univ Chem 11(1):18–30
32. Wang B, Liang F, Wang Y, Peng K (2016) TMS light metals 2016, pp 429–431
33. Cao D, Ma Y, Shi Z, Xu J, Hu X, Wang Z (2019) Corros Sci 156:32–43. https://doi.org/10.1016/j.corsci.2019.05.012
34. Simakov D, Gusev A (2015) Patent RF 2570149
35. Chernov R, Perepelitsa A, Antishko A (1975) Appl Chem J 48(3):634–635 (in Russian)
36. Istomin S, Kulikov B, Kustov V (1992) Tsvetnye Met (Non-ferrous Met) 1:32–35 (in Russian)
37. Istomin S, Golovnykh N, Bychinsky V, Kulikov B (2003) Tsvetnye Met (Non-ferrous Met) 2:55–58 (in Russian)
38. Haarberg G (2020) ECS Trans 98(10):181–187
39. Belyaev A (1944) Metallurgiy alegkikh metallov (Metallurgy of light metals). Metallurgizdat, Moscow, 543 p (in Russian)
40. Nerubashchenko V, Krymov A, Voleinik V (1977) Tsvetnye Met (Non-ferrous Met) 7:29–31 (in Russian)
41. Nerubashchenko V, Voleinik V, Krymov A (1978) Tsvetnye Met (Non-ferrous Met) 3:49–51 (in Russian)
42. Nerubashchenko V, Krymov A, Bugaenko V (1983) Izv VUZov Non-ferrous Metall 6:49–51 (in Russian)
43. Grjotheim K, Matiasovski K (1983) Aluminium 59(9):682–693

Index

D. Pruttskov et al., *Electrolytic Production of Al–Si Alloys*, Monographs in Electrochemistry, https://doi.org/10.1007/978-3-031-29249-1

SPRINGER NATURE

GPSR Compliance

The European Union's (EU) General Product Safety Regulation (GPSR) is a set of rules that requires consumer products to be safe and our obligations to ensure this.

If you have any concerns about our products, you can contact us on ProductSafety@springernature.com

In case Publisher is established outside the EU, the EU authorized representative is:

Springer Nature Customer Service Center GmbH
Europaplatz 3
69115 Heidelberg, Germany

The manufacturer's authorised representative in the EU is Springer Nature Customer Service Centre GmbH, Europaplatz 3, 69115 Heidelberg, Germany. If you have any concerns regarding our products, please contact ProductSafety@springernature.com

Printed and bound by CPI Group (UK) Ltd, Croydon, CR0 4YY
01/10/2026
02225882-0004